HARD WORK

Defining Physical Work Performance Requirements

Brian J. Sharkey, PhD
USDA Forest Service
Technology and Development Center
Missoula, MT

Paul O. Davis, PhD
First Responder Institute
Burtonsville, MD

Human Kinetics

Library of Congress Cataloging-in-Publication Data

Sharkey, Brian J.
  Hard work : defining physical work performance requirements / Brian J.
Sharkey, Paul O. Davis.
      p. ; cm.
  Includes bibliographical references and index.
  ISBN-13: 978-0-7360-6536-8 (hard cover)
  ISBN-10: 0-7360-6536-9 (hard cover)
  1. Work capacity evaluation. I. Davis, Paul O. II. Title.
  [DNLM: 1. Work--physiology. 2. Biomechanics. 3. Exertion. 4. Work Capacity Evaluation. WE 103 S531h 2008]
  RC963.4.S43 2008
  612'.042--dc22
                                       2007040245

ISBN-10: 0-7360-6536-9
ISBN-13: 978-0-7360-6536-8

The Web addresses cited in this text were current as of September 4, 2007, unless otherwise noted.

Acquisitions Editor: Michael S. Bahrke, PhD; **Developmental Editor:** Rebecca Johnson; **Managing Editors:** Heather M. Tanner and Maureen Eckstein; **Assistant Editor:** Christine Bryant Cohen; **Copyeditor:** Patsy Fortney; **Proofreader:** Erin Cler; **Indexer:** Gerry Lynn Shipe; **Permission Manager:** Carly Breeding; **Graphic Designer:** Fred Starbird; **Graphic Artist:** Dawn Sills; **Cover Designer:** Keith Blomberg; **Photographer (cover):** © Martha Ellis; **Photo Asset Manager:** Laura Fitch; **Photo Office Assistant:** Jason Allen; **Art Manager:** Kelly Hendren; **Associate Art Manager:** Alan L. Wilborn; **Illustrator:** Mic Greenberg; **Printer:** Thomson-Shore, Inc.

Printed in the United States of America   10   9   8   7   6   5   4   3   2   1

Human Kinetics
Web site: www.HumanKinetics.com

United States: Human Kinetics
P.O. Box 5076
Champaign, IL 61825-5076
800-747-4457
e-mail: humank@hkusa.com

Canada: Human Kinetics
475 Devonshire Road Unit 100
Windsor, ON N8Y 2L5
800-465-7301 (in Canada only)
e-mail: orders@hkcanada.com

Europe: Human Kinetics
107 Bradford Road
Stanningley
Leeds LS28 6AT, United Kingdom
+44 (0) 113 255 5665
e-mail: hk@hkeurope.com

Australia: Human Kinetics
57A Price Avenue
Lower Mitcham, South Australia 5062
08 8372 0999
e-mail: info@hkaustralia.com

New Zealand: Human Kinetics
Division of Sports Distributors NZ Ltd.
P.O. Box 300 226 Albany
North Shore City
Auckland
0064 9 448 1207
e-mail: info@humankinetics.co.nz

To the men and women who strive to acquire
and maintain the fitness they need to get up each day,
put on their boots, and go forth to perform hard work.

Contents

Preface   ix
Acknowledgments   x

PART I
The Job and the Worker 1

Chapter 1   Physically Demanding Occupations. 3
Hard Work   4
Work Requirements   8
Work Capacity   9
Summary   11

Chapter 2   The Worker 13
Physical Characteristics of Workers   14
Aerobic Fitness   14
Muscular Fitness   20
Demographic Trends   23
Summary   25

Chapter 3   Employment Opportunity 27
Employment Laws   28
Uniform Guidelines on Employee Selection Procedures   30
Summary   32

PART II
Test Development and Validation . . .35

Chapter 4   Job-Related Tests. 37
Why Test?   37
Testing and Legal Issues   39
Test Development   40
Summary   47

Chapter 5   Test Validation **49**

Content Validity   49

Criterion-Related Validity   50

Construct Validity   51

Validation Options   51

Reliability and Cross-Validation   52

Bona Fide Occupational Qualification   53

Absolute Standards   53

Test Standards   54

Suboptimal Selection Procedures   57

Summary   57

Chapter 6   Test Implementation **59**

Seeking Professional Assistance   60

Personnel Issues   61

Implementation Strategies   63

Summary   64

PART III
Employee Selection Practices 65

Chapter 7   Testing New Employees **67**

Medical Standards   68

Safety Considerations   71

Instructions   72

Recruitment   72

Pretest Training   73

Retesting   74

Work Hardening   74

Summary   75

Chapter 8   Testing Incumbent Employees **77**

Age and Performance   78

Physiological Age   79

Periodic Testing   80

Providing Adequate Notice   82

Medical Standards and Safety   82

Fitness Training   83

Test Results   83

Summary   85

Chapter 9   Program Evaluation. 87

Evaluation   88

Surveillance System   90

Analysis: Quantitative and Qualitative Information   91

Cost-Benefit Analysis   92

Reporting Results   93

Summary   93

PART IV
Employee Health, Physiology, and Performance 95

Chapter 10   Employee Health 97

Developing Medical Standards   98

Employee Health Programs   102

Costs and Benefits of Employee Health Programs   106

Summary   107

Chapter 11   Physiology of Work 109

Muscle Fibers   110

Muscle Contractions   110

Energy Sources   112

Oxygen and Energy   115

Supply and Support Systems   116

Training Effect   119

Summary   120

Chapter 12   Job-Related Fitness. 121

Job-Related Fitness Programs   122

Aerobic Fitness   123

Muscular Fitness   128

Core Training   130

Periodizing the Training Plan   131

Body Composition   131

Program Issues   132

Risks and Benefits   134

Summary   136

PART V

Job-Related Issues 137

Chapter 13 Environmental Impacts 139

Heat Stress and Heat Disorders 140
Preventing Heat Disorders 141
Cold Conditions 146
Altitude Acclimatization 147
Summary 148

Chapter 14 Respiratory Protection 151

Respiratory Hazards 152
Respirator Selection 152
Respiratory Protection Program 153
Medical Evaluation 154
Work Performance 155
APRs and Women 157
Summary 157

Chapter 15 Lifting Guidelines 159

Lifting Standards 160
NIOSH Lifting Equation 161
Selection and Training 164
Summary 166

Chapter 16 Legal Issues 167

Legal Challenges 168
Unintended Consequences 170
Court Decisions 171
Alternative to Litigation 174
Summary 177

Appendix A   Physical Ability Testing: Conflicts, Conundrums, and Consequences 179
Appendix B   Functional Testing in Selection, Placement, and Return-to-Work 191
Appendix C   A Chief's Personal Narrative 195
Appendix D   Work Output as a Function of Selectivity in Hiring 199
Appendix E   Biochemical Evaluation of Workplace Stress 207
Appendix F   Physical Fitness Policy 211
Appendix G   Standard Operating Procedures 213
Appendix H   Essential Functions of Law Enforcement 217
Appendix I   Assessment of Performance During Manual Timber Harvesting 221
References 223
Index 229
About the Authors 237

Preface

Since *Homo sapiens* emerged from the heart of Africa, men and women have toiled and struggled to survive, prosper, and build. The sweat of the brow was the consequence of this effort. With the domestication of beasts of burden, some of the heavy lifting was outsourced. But it was not until the industrial age and the advent of the steam engine, the internal combustion engine, and electrification that profound changes in productivity occurred. Engineering the toil out of work has had a major impact on how we now relate to the performance of physically demanding occupations.

With the evolution of corporations, labor organizations, and the legal system came an acceptance of the concept of employment rights. This in turn necessitated the creation of complex laws to address issues of opportunity, adverse or disparate treatment, and discrimination. This book focuses on hard work, physically demanding occupations that require strength, stamina, or both. The perspective is that of the work physiologist (also called occupational or applied physiologist or exercise scientist), who studies the physiological responses to arduous work; constructs tests of work capacity; designs training programs to enhance performance; and provides nutrition, hydration, and other support. Although many exercise scientists focus on the application of research to performance in sport, the authors serve workers who toil daily at jobs that require considerable strength and stamina. At their physical best, workers in this cohort could be viewed as occupational athletes.

The athletic model presented in this book is not common in the workplace. Some occupations do little to screen for suitable candidates. Those that do test recruits are not likely to require ongoing fitness training or annual tests of work capacity.

The authors of this book have collectively spent over 70 years studying physically demanding work and the factors associated with performance and health. We have studied workers in environments ranging from the jungles of Panama to the Rocky Mountain wilderness, from the heat of the Mojave desert to winter in the Sierras. We have studied structural and wildland firefighters, law enforcement officers, and military personnel. We've conducted studies in the laboratory and in the field and have published papers and reported our findings at meetings throughout the world. Finally, we have served as expert witnesses in cases related to work capacity and performance.

This book attempts to dispel misconceptions, simplify the complex, and provide an approach for making intelligent and informed employment decisions that will result in a safer, healthier, and more productive workforce. Part I examines the job, the worker, and the legal aspects of employment. Part II focuses on test development and validation, and part III shows how tests are used in employee selection. Part IV considers employee health and performance, and part V discusses job-related issues such as environmental concerns, respiratory protection, lifting guidelines, and legal issues. Throughout the book you will find case studies that illustrate the good, the bad, and the ugly sides of employee selection and physical maintenance in demanding occupations. The selected appendixes in the back of the book offer varying perspectives on testing, job-related fitness, and work performance, as well as on testing and fitness policies and procedures.

The book is intended for a wide-ranging audience, including work (occupational) physiologists and graduate students; physicians in occupational medicine; managers and employees in physically demanding occupations; city and agency personnel specialists; judges, lawyers, and expert witnesses in legal cases involving employment in physically demanding occupations; personnel in state and federal labor and justice departments; and those who write and interpret employment laws. We look forward to hearing your reactions to our endeavor to write the first book of its kind in the field.

Acknowledgments

We thank the hardworking employees, studied in the laboratory and in the field, who have taught us the meaning of hard work. We credit our colleagues and coworkers who have helped shape and focus our understanding of physically demanding occupations. We acknowledge the contributions of Norm Henderson, PhD; Delia Roberts, PhD; Bill Gillman, PhD; Lee Cunningham, PhD; Paul DiVico, EdD; Gerald Villella, LLD; Saul Krenzel, LLD; Wayne Schmidt; Douglas MacDonald; John Key; John LeCuyer; and Don Oliver to this effort. And we thank publishers Rainer Martens and Julie Simon for their confidence in our ability to communicate the importance of job-related fitness to health, safety, and performance. Last but not least, we appreciate the efforts of Michael Bahrke, Rebecca Johnson, Heather Tanner, Maureen Eckstein, Christine Cohen, and the rest of the talented crew at Human Kinetics, who helped to transform our manuscript into a book.

PART I
The Job and the Worker

© Tyler Stableford/Stone/Getty Images

In order that people may be happy in their work, these three things are needed: They must be fit for it. They must not do too much of it. And they must have a sense of success in it.

John Ruskin

Many years ago, the primary source of power for the production of useful work was the contractions of muscles, both human and animal. People devised ways to augment muscle power with the ingenious use of wind and water, but it was not until the 19th century that mechanization began to reduce the need for muscular work. Machines were devised to supplement or replace human effort in the factory. Later, robots, computers, and other devices replaced human muscle power. Today, few men and women are required to engage in arduous muscular effort at work.

Much of the credit for the reduction in physical labor must go to the inventors and engineers whose attempts at mechanization and, more recently, computers and robotics have made work relatively effortless. Some credit also is due to specialists in ergonomics, the scientific study of work: Engineers, occupational physiologists, psychologists, and physicians study men and women in their working environments with the goal of adapting the job to the ability of the average worker. At the same time, laborsaving devices have eliminated the need for muscular work at home, and the automobile makes the task of getting to and from work physically effortless. The consequences of these trends are obvious: The average worker is incapable of delivering a full day's effort in a physically demanding job, and degenerative diseases associated with inactivity and overweight, such as heart disease—the number one killer in the United States—are epidemic.

And yet, physically demanding jobs still exist in mining, forestry, agriculture, construction, and public safety. In firefighting and law enforcement, the physical demands of the job have public safety and property implications. In wildland firefighting, the ability to escape to a safety zone is highly correlated with a worker's aerobic fitness (Ruby et al. 2002). In the military, where every service member has the potential to be involved in combat, the physical demands of the job cannot be ignored. Clearly, a substantial number of people continue to work in physically demanding occupations.

In sport, athletes undergo training to help them meet the demands of the event, but few employers offer training to help workers meet the demands of work. Perhaps it is time for a change; perhaps workers could benefit by measuring up to job requirements. When it isn't possible to adapt the job to the worker, perhaps the worker can adapt to the job. A few companies have viewed their workers as occupational athletes and have provided training to help them excel in the workplace. One of the goals of this book is to help employers identify capable people and provide the training those people need to sustain that capability.

In part I, chapters 1 and 2 define physically demanding occupations and the characteristics of workers available for employment. Chapter 3 considers the legal aspects of employment, including age, gender, race, disability, and other employment issues.

Chapter 1

Physically Demanding Occupations

© Alan Thornton/Stone+/ Getty Images

As a teenager, Brian Sharkey, one of this book's authors, had lots of jobs, including yard work, painting, and farm work. When he reached the age of 15, he lied about his age and got a job on a construction crew: That was hard work! He was up early so he could get to the construction office before 7 a.m., where he and the crew loaded the truck, climbed aboard, and rode for up to an hour to the construction site. At 8 a.m., when they went on the clock, they began the job of building concrete foundations in a housing development. Part of the crew prepared the next site while others stripped concrete forms from yesterday's pour. The heavy forms were made of a sheet of 3/4-inch 4 × 8 foot plywood nailed to a 2 × 4 inch frame. They were lifted out of the foundation hole and carried to the next site.

By break time on his first day, Brian was suffering from the heat and humidity. Even though he was an endurance athlete, it was all he could do to keep working. Brian's first paycheck was less than the amount he expected. The foreman said he hadn't pulled his weight during the first week of work. Instead of quitting, he adjusted to the heat, the humidity, and the physical demands of the job. He received full pay for the rest of that summer and several more summers, until he took a job as an apprentice butcher.

Paul Davis, coauthor of this book, was also attracted by the pay for construction work. He left a job as carry-out clerk in a grocery store to work as a laborer. His first assignment was to burrow under a sagging foundation to insert a beam and jack to prop up the wall. The dust was oppressive, so he improvised with a wet bandana (this was before the days of safety regulations). He suffered through severe blisters but eventually accomplished the task. This was the first of many construction tasks for Paul, such as digging footings, unloading cinderblocks, or moving 5 yards (4.6 m) of gravel that had been dumped in the wrong location.

This book describes the absolute requirements of hard work and teaches employers how to select, train, and maintain capable employees. We cannot address all the physically demanding occupations that exist, such as sheep shearers, those impressive Australian and New Zealand athletes who shear several hundred sheep per day; laborers who carry immense loads, sometimes on their heads; or South African miners who work in hot, confined conditions. There are many occupational athletes in the world, but this book focuses on the major occupational and military groups in North America.

This chapter will help you do the following:

- Identify the components of physically demanding occupations.
- Determine the work requirements associated with physically demanding occupations.
- Define the concept of work capacity.

Hard Work

Certain jobs require strength and endurance at least some of the time. Workers in heavy industry, construction, agriculture, mining, forestry, firefighting, law enforcement, and the military are often required to engage in strenuous effort.

Work is considered physically demanding when it requires high energy expenditure, intense muscular effort, or a combination of the two. Other components can make work difficult and stressful, such as the vigilance required of air traffic controllers, the balance and agility required of steelworkers, or the endless repetition of the assembly line. In this section we discuss the two components that define physically demanding occupations: energy expenditure and muscular effort.

Energy Expenditure

To perform work, the body requires the energy that is released during the metabolism of fat and carbohydrate. This process requires oxygen; the tougher the job is, the more energy and oxygen are needed. When oxygen and energy needs are light, such as for office duties, work performance isn't related to the ability to take in, transport, and use oxygen (aerobic fitness). But when oxygen and energy needs are moderate or high (over 5 kcal per minute), work production relates directly to the ability to sustain aerobic energy production (see table 1.1).

The energy expenditure of a job can be measured in the field using indirect calorimetry. The worker breathes through a valve that directs expired air into a device that measures the composition of the air (oxygen and carbon dioxide) and the volume of air expired during the task. A computer records the data, adjusts for temperature and barometric pressure, and calculates the amount of oxygen consumed. For example, say that in one minute a worker expires 40 liters of air with an oxygen content of 18.43 percent. Because atmospheric air

Table 1.1 Physical Work Classification: Energy Expenditure

Classification	Caloric cost	Oxygen cost	Examples
Light	<2.5 kcal/min	<0.5 L/min	Desk work
Moderate	2.5-5	0.5-1.0	Chores, repairs, walking
Heavy	5-7.5	1.0-1.5	Chainsawing, shoveling
Very heavy	7.5-10	1.5-2.0	Chopping wood, digging
Extremely heavy	>10	>2.0	Hiking uphill with pack, jogging

Adapted from U.S. Department of Labor, 1968, *Dictionary of occupational titles* (Washington, D.C.: U.S. Government Printing Office) and P.O. Astrand et al., 2003, *Textbook of work physiology*, 4th ed. (Champaign, IL: Human Kinetics).

is 20.93 percent oxygen, the worker has used 2.5 percent of oxygen × 40 liters per minute, or 1.0 liter of oxygen per minute. Metabolic chamber studies have shown that 1 liter of oxygen is equivalent to 5 kilocalories of energy expended. According to table 1.1, this job is between moderate and heavy work. It would be relatively easy to do for an hour or two, but demanding if done all day long.

Moderate levels of work can be demanding when done for long periods and when environmental factors such as heat and humidity increase the physiological stress on the worker. Heavy and very heavy work place a greater burden on the worker, necessitating more frequent rest breaks. Extremely heavy work (above 10 kcal or 2 L of oxygen) occurs in the military, law enforcement, and structural firefighting. This work can be sustained for prolonged periods by highly trained workers, occupational athletes specifically conditioned for the task, but can be done only intermittently by those with lower levels of fitness (see figure 1.1).

When hard work is required, people with a low level of aerobic fitness are able to sustain only 25 percent of capacity for eight hours. Those of average fitness can work at about 33 percent, whereas those with above-average fitness can maintain 40 percent of their capacity for eight hours. Only highly conditioned and motivated workers can sustain levels as high as 50 percent of their aerobic fitness level for eight hours.

Some occupations require that workers engage in bouts of extremely heavy work interspersed with relative rest. Police officers spend hours behind the wheel of a cruiser, but can be called on for an all-out effort to catch and subdue a suspect. Structural firefighters work hard to extinguish a fire, then recover while their self-contained breathing apparatus (SCBA) is fitted with a fresh bottle of air.

When the required energy expenditure is extremely high, it may exceed the aerobic capacity of some workers. When this happens, workers use energy from nonoxidative, or anaerobic, metabolism. Anaerobic energy production is very inefficient, producing 3 units of energy (adenosine triphosphate, or ATP) from one molecule of

Energy expenditure of a job can be measured in the field using a calorimeter.

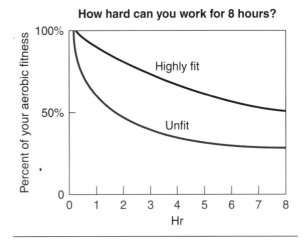

How hard can you work for 8 hours?

Figure 1.1 Fitness and work capacity.

Reprinted, by permission, from B.J. Sharkey, 2002, *Fitness and health*, 5th ed. (Champaign, IL: Human Kinetics), 310.

glucose, whereas oxidative, or aerobic, energy production produces 36 units of ATP from the same amount of glucose. For this reason, anaerobic work can be sustained only for brief periods. Chapter 2 has more information about the contribution of aerobic fitness in physically demanding occupations, and chapter 11 provides more details about aerobic and anaerobic metabolism.

Muscular Fitness

Muscular fitness consists of strength, muscular endurance, and power. Strength is the ability to lift heavy objects; muscular endurance is the ability to do repetitive contractions with lesser weights; and power is the ability to do work (force × distance) rapidly (force × distance / time). Dynamic muscular strength is clearly related to work capacity when the work requires the lifting of very heavy loads or the use of heavy tools (see table 1.2).

Construction workers engage in heavy work when they lift and carry lumber and concrete bags weighing up to 100 pounds (45.5 kg) and lift heavy frame walls. Structural firefighters wear personal protective clothing and breathing apparatuses (SCBAs) that may weigh as much as 50 pounds (22.7 kg). Then they lift hoses and tools, climb steps or ladders, and fight the fire. Many work tasks require more endurance than strength. If a worker has the strength to accomplish a task, physical conditioning should focus on developing muscular endurance and aerobic fitness. Those with inadequate strength must engage in strength training.

Power, the rate of doing work, is a combination of force (F) and speed (distance / time). It can be useful in law enforcement to subdue a suspect and in firefighting to chop through a roof, pull a charged hose, or rescue a victim. Inadequate power can make some tasks difficult.

Lean Body Weight

Lean body weight (LBW) equals body weight minus fat weight. When considerable muscular strength is required, the person with a higher lean body weight (LBW), or fat-free mass (FFM), is more likely to excel than someone with a lower LBW. The LBW indicates how much muscle is brought to the job. Body weight alone doesn't tell the whole story

Table 1.2   Physical Work Classification: Muscular

Work classification	Lifting lb (kg)	Carrying lb (kg)
Very light	Up to 10 (4.5)	Small objects
Light	Up to 20 (9.1)	Up to 10 (4.5)
Medium	Up to 50 (22.7)	Up to 25 (11.3)
Heavy	Up to 100 (45)	Up to 50 (22.7)
Very heavy	Over 100 (45)	Over 50 (22.7)

Adapted from U.S. Department of Labor, 1968, *Dictionary of occupational titles* (Washington, D.C.: U.S. Government Printing Office) and P.O. Astrand et al., 2003, *Textbook of work physiology*, 4th ed. (Champaign, IL: Human Kinetics).

©iStockphoto.com/Ann Marie Kurtz

Construction workers engage in heavy work when they lift and carry lumber and concrete bags weighing up to 100 pounds (45.5 kg).

about a person's body composition. The percentage of body fat can also be misleading: 20 percent fat sounds high, unless it is associated with 200 pounds (90.7 kg) of lean weight.

In a study of wildland firefighters, work capacity was highly related to LBW (Sharkey 1981). Correlations as high as .80 have been reported for the relationship between military material handling and LBW (Hodgdon 1992). LBW has relevance for women applying for positions that require hard work. Because women typically have a greater percentage of body fat than men, men and women of the same weight will not have the same LBW. To have a similar level of LBW as a male applicant, a female applicant will have to have less than the average percentage of body fat or weigh more than the male applicant.

Energy and Muscular Requirements

For repeated lifting, as in moving loads or working with hand tools, energy and muscular requirements combine to set the limits of work capacity. Workers with a high level of aerobic fitness can work at a higher work rate than those with lower levels; they can do more contractions per minute. Because the load is a lower percentage of their maximal strength, stronger workers can lift more with each repetition (see figure 1.2).

Combinations of work rate and percentage of maximal strength that fall to the right of the line (in figure 1.2) that represents one's level of aerobic fitness cannot be sustained for a full working day. Highly fit workers (aerobic fitness = 55 ml · kg^{-1} · min^{-1}) can produce more by working at a higher rate. Stronger workers can lift more with each contraction. The ideal combination includes a balance of aerobic fitness and strength adequate to the task.

The muscular demands of work can be assessed by weighing the loads that must be lifted and carried and by measuring the work rate of successful workers. We will say much more about lifting requirements in chapter 15, Lifting Guidelines.

Energy requirements of many occupational tasks have been studied. New or unique tasks can be studied in the field using indirect calorimetry. Table 1.3 summarizes the costs of a number of tasks. These are average values from past studies. Workers can raise the energy cost above the listed value by working very hard or work well below the value by taking their time. Note that these values are for a 150-pound (68 kg) worker and should be adjusted 10 percent for each 15 pounds (6.8 kg) above or below 150 pounds (68 kg) (see table 1.3 footnote).

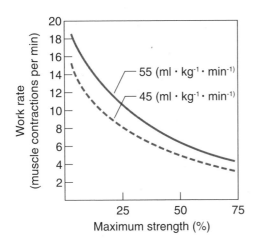

Figure 1.2 Strength, aerobic fitness, and work rate.

Reprinted, by permission, from B.J. Sharkey and S.E. Gaskill, 2007, *Fitness and health*, 6th ed. (Champaign, IL: Human Kinetics), 321.

⅏⅏⅏⅏⅏⅏⅏ HOW MUCH ⅏⅏⅏⅏⅏⅏⅏⅏ STRENGTH IS NECESSARY?

Generally speaking, for prolonged work, the average load in repetitive lifting should not exceed 20 percent of the worker's maximal strength in that movement: Strength should be at least five times greater than the load regularly lifted on the job. If the job requires daylong work with a shovel that weighs 10 pounds (4.5 kg) when loaded, the worker should possess at least 50 pounds (22.7 kg) of dynamic muscular strength in the arms, shoulders, and back muscles used to perform the task. Once the worker has more than the minimal strength required, further increases in work capacity can be achieved by increasing muscular endurance and aerobic fitness. If the job involves only occasional lifts of very heavy loads (e.g., 100 pounds [45.5 kg]), the worker can succeed with 100 pounds of strength plus a margin for safety. The greater the margin is, the less risk there is of injury, because the risk of certain injuries (e.g., low back) increases with fatigue. Of course, the worker has to use good lifting techniques to minimize the likelihood of injury.

Table 1.3   Energy Requirements of Work Tasks

Work task	Kcal/min	Work task	Kcal/min
LIGHT WORK		**MEDIUM WORK**	
Desk work	2.5	Paving roads	5.0
Standing, light activity	2.6	Gardening, weeding	5.6
Driving	2.8	Walk 3.5 mph (5.6 km/h) (road/field)	5.6, 7.0
Washing clothes	3.1	Stacking lumber	5.8
Walking indoors	3.1	Chainsawing	6.2
Making bed	3.4	Laying stone	6.3
Drive motorcycle	3.4	Using pick and shovel	6.7
Metalworking	3.5	Shoveling (miners)	6.8
House painting	3.5	Walking downstairs	7.1
Walk downhill 2.5 mph (-5, 10, 15% grade)	3.5, 3.7, 4.3	**HEAVY WORK**	
Cleaning windows	3.7	Shoveling snow	7.5
Carpentry	3.8	Chopping wood	7.5
Farm chores	3.8	Hike with 45 lb (20kg) pack	7.5
Sweep floors	3.9	Crosscut sawing	7.5, 10
Plastering walls	4.1	Walking uphill 3.5 mph (5.6 km/h) (+5, 10, 15% grade)	8, 11, 15
Auto repair	4.2	Tree felling (ax)	8.5, 12
Ironing	4.2	Gardening, digging	8.6
Raking, hoeing	4.7	**VERY HEAVY WORK**	
Mix cement	4.7	Walking upstairs	10+
Mopping floors	4.9	Jogging	10

Energy cost depends on efficiency and body size. Add 10% for each 15 lb (6.8 kg) over 150 lb (68 kg); subtract 10% for each 15 lb (6.8 kg) below 150 lb (68 kg).

Adapted from Human Performance Laboratory, University of Montana, 1964-present, R. Passmore and J. Durnin, 1955, "Human energy expenditure," *Physiological Review* 35: 801-824, and E. Roth, 1968, *Compendium of human responses to the aerospace environment: Volume III* (Washington, D.C., National Aeronautics and Space Administration).

Work Requirements

Not all work is physically demanding. Many factory jobs have been made less difficult with mechanical assists to reduce energy and muscular demands. Forestry, once one of the most demanding of occupations, is now made easier with mechanized tree harvesters that cut, delimb, and stack trees. The backbreaking work of the longshoreman or stevedore has been eliminated with cranes and cargo boxes. But mechanized aids are not available for all occupations, such as firefighting, construction, and the military.

The energy and muscular requirements of hard work can be determined from published studies or estimated in the field. Energy expenditure can be estimated from working heart rates. Muscular requirements can be estimated by observing the loads lifted and carried, and work rates can be estimated by recording the number of lifts per minute. Table 1.4 lists the energy and muscular requirements for selected occupations. Because few occupations sustain a single work rate, a range of values is provided.

Those in the U.S. Army and the Marine Corps are more likely than personnel in other branches of the military to carry heavy packs and weapons for extended periods. Those in the U.S. Navy and Air Force have better access to energy-saving devices to lift and carry loads. However, in a time of war, U.S. Department of Defense policy states that every uniformed service member must be combat ready, that is, able to perform combat requirements such as negotiating the battlefield, reacting to combat situations, and controlling or evacuating crowds.

Table 1.4 Occupational Work Classifications

Occupation	Energy (kcal/min)	Muscular*	Comments
Farm labor	4.0->7.5	Medium	Pick, carry, load
Construction	4.0-10	Heavy	Hand tools, lift, carry
Mining	4-7.5	Light-heavy	Pick, shovel
Law enforcement	2.5-10+	Very light-very heavy	Run, subdue
FIREFIGHTING			
Structural	7.5-10	Medium-very heavy	Climb, chop, lift, carry
Wildland	6.5-10	Heavy-very heavy	Dig line, carry hose
MILITARY			
Army, Marines	2.5-10+	Medium-very heavy	Hike, climb, carry
Navy, Air Force	2.5-7.5	Medium-heavy	Lift, carry, load

*See lifting and carrying loads in table 1.2.

Work Capacity

Work capacity is defined as workers' ability to accomplish production goals without undue fatigue and without becoming a hazard to themselves or coworkers and burdening the employee health plan. It can also be viewed as sustainable work capacity, the level of effort that can be sustained throughout the work shift.

CASE STUDY
Wildland Firefighters

Wildland firefighters are called on to fight fires in forests and grasslands. Although some fires can be extinguished in hours, others take days, weeks, or even months, such as the 1988 fires in Yellowstone National Park. For this reason firefighters need the endurance to work up to 16 hours a day for several weeks at a time. Most large fires occur in hot, dry weather and in terrain that is rough and steep. In 1962 officials from the U.S. Forest Service Technology and Development Center and the University of Montana Human Performance Laboratory signed a cooperative research agreement to conduct studies on wildland firefighters. Initial research consisted of studies of the energy, thermal, and cardiorespiratory demands of firefighting tasks.

Douglas bags were used to collect expired air samples. The contents were analyzed for oxygen and carbon dioxide, and the volume was measured with a gas meter. Devices that metered expired air and collected a sample for analysis were also used. In the 1970s expired air collected in meteorological balloons was analyzed in a computerized device. In the 1990s lightweight computerized analyzers weighing 1 kilogram (2.2 lb) were used for field studies. It is now possible to telemeter information from the breathing valve to a nearby computer.

The most recent analysis used doubly-labeled water, the gold standard for energy expenditure studies. Using a small cocktail with the isotopes of hydrogen (8H) and oxygen (18O), this expensive but accurate technique allowed researchers to calculate energy expenditure over a number of days. The results demonstrated the high level of firefighter energy expenditure, sometimes exceeding 6,000 kilocalories per day (Ruby et al. 2002). The recent analysis confirmed earlier findings that wildland firefighters burn about 7.5 kilocalories per minute, and over 400 kilocalories per hour, for 12- to 16-hour shifts, values seen in few occupations. Measurements of the energy expenditures of wildland firefighters in Canada and bush firefighters in Australia were similar to those of firefighters in the United

(continued)

(continued)

States (Budd et al. 1997; Docherty, McFadyen, and Sleivert 1992).

In addition to energy expenditure, the muscular demands of firefighting tasks have been studied. (Budd et al. 1997; Docherty et al. 1992; Sharkey 1997b). Firefighters carry tools and wear field packs that weigh at least 25 pounds (11.4 kg) with food, water, and a fire shelter (a reflective tent for emergency use). Firefighters lift and carry 65-pound (29.5 kg) pumps. They carry and deploy hoses and move them when charged with water. And they sometimes carry backpack pumps weighing 45 pounds (20.5 kg). Laboratory and field studies, such as The National Wildfire Coordinating Group's Wildland Firefighter Health and Safety Reports by Sharkey (2000-2006), have looked at uniforms, tools, nutrition, hydration, and issues related to firefighter health, safety, and performance. For more information on these reports, see www.fs.fed.us/fire/safety/h_s_rpts/.

A study of wildland firefighters clearly shows the link between fitness and work capacity (Sharkey 1981). Fit workers do more work with less fatigue and still have a reserve to meet unforeseen emergencies. Fit workers are better able to reach a safety zone in an emergency (Ruby et al. 2002). Fit workers acclimatize to the heat faster than less fit workers, perform better in the heat, and recover more quickly from adverse conditions, such as long shifts and reduced rest. In some tasks the fit worker is able to produce several times more than a marginally fit worker (Sharkey 1981). Aerobic fitness is the most important factor in work capacity for wildland firefighters. From 1975 to 1998 U.S. federal agencies used an aerobic fitness test to determine work capacity. The initial test was revised in 1998; the new tests are discussed in chapter 5.

A Douglas bag in use.

Photo courtesy of the USDA Forest Service, Missoula Technology and Development Center

The National Wildfire Coordinating Group (NWCG) coordinates five U.S. federal agencies and state forestry agencies through the National Association of State Foresters. They have developed standards for workers involved in a number of arduous, moderate, and light duties related to fire suppression (NWCG 310-1). They include the following:

Arduous Work

"Duties involve field work requiring physical performance calling for above-average endurance and superior conditioning. These duties may include an occasional demand for extraordinarily strenuous activities in emergencies under adverse environmental conditions and over extended periods of time. Requirements include running, walking, climbing, jumping, twisting, bending, and lifting more than 50 pounds; the pace of work typically is set by the emergency condition." (This description applies to wildland firefighters and those with demanding supervisory roles.)

Moderate Work

"Duties involve field work requiring complete control of all physical faculties and may include considerable walking over irregular ground, standing for long periods of time, lifting 25 to 50 pounds, climbing, bending, stooping, squatting, twisting, and reaching. Occasional demands may be required for moderately strenuous activities in emergencies over long periods of time. Individuals usually set their own work pace." (This description applies to those with a range of field activities, such as a safety officer.)

Light Work

"Duties mainly involve office type work with occasional field activity characterized by light physical exertion requiring basic good health. Activities may include climbing stairs, standing, operating a vehicle, and long hours of work, as well as some bending, stooping, or light lifting. Individuals almost always can govern the extent and pace of their physical activity." (This description applies to individuals with responsibilities in the fire camp.)

The arduous and moderate categories fit our description of physically demanding occupations. Each year in the United States more than 25,000 people qualify to work on fires. They are joined by state, county and rural firefighters, and even members of the National Guard in bad fire seasons.

Work capacity is the product of a number of factors, including natural endowment, skill, nutrition, aerobic and muscular fitness, intelligence, experience, acclimatization, and lean body weight (Sharkey 1997b).

Aerobic or muscular fitness, acclimatization to heat or altitude, and even skill and experience do not ensure work output. A worker may rate highly in all of these categories, but fail to produce adequately because of a lack of motivation. On the other hand, even the most highly motivated workers will fail if they lack strength or endurance, ignore the need to acclimatize to a hot environment, or lack the physical skills required for the job.

Summary

This chapter discussed the components of hard work, energy expenditure, and muscular demands. Energy expenditure can be moderate and prolonged or extremely heavy but brief. As energy expenditure increases in intensity, the duration must be reduced. In the workplace, that means more rest breaks. Muscular demands are defined by the lifting and carrying requirements of the job. When more than 50 pounds (22.7 kg) must be lifted or carried, that is hard work. High energy expenditure and intense muscular demands can combine to make work physically demanding.

Examples of physically demanding occupations include farm labor, construction, mining, law enforcement, firefighting, and the military. Some demanding occupations have been made easier with mechanical assists. But in others, where mechanization is not available, the job continues to be hard, too hard for the average person.

Work capacity defines the requirements of a given job. Hard work usually involves a high level of aerobic and muscular fitness. Those who meet the requirements are able to perform while maintaining a margin of safety and an ability to respond to emergencies. Those who lack these requirements are more likely to be injured, develop overuse injuries, incur workers' compensation costs, and become ill and burden employee health plans.

Chapter 2
The Worker

Photo courtesy of the USDA Forest Service, Missoula Technology and Development Center

Tara was a graduate research assistant when she decided to become a smoke jumper, a wildland firefighter who parachutes into wilderness fires. She had served two years on a fire crew and wanted to become a member of this elite firefighting organization. At that time only a few women had made it through training to become jumpers. She was an endurance athlete who barely weighed enough to qualify for training, so she designed and carried out a weight training program to gain strength and lean body weight. After several months of serious weight training, she successfully completed the qualifying test, which included a 1.5-mile (2.4 km) run in 11 minutes and pull-ups, push-ups, and training, including the pack-out—a 3-mile (4.8 km) hike with a 110-pound (50 kg) pack. This load, which exceeded her lean body weight, simulates smoke jumper gear—including a parachute, padded jump suit, and helmet—which jumpers must pack out of the back country. This determined woman wanted to be accepted on her merits as a qualified worker, capable of excelling in a very demanding job.

Where do we find people with the strength and stamina needed to engage in physically demanding occupations? Are U.S. citizens up to the challenges of hard work? Politicians tell us that there are jobs Americans won't do, but that isn't true. Americans will do hard work as long as the job is rewarding and the compensation is adequate for the effort. The military and firefighting vocations demonstrate that physically demanding jobs do attract exceptional workers. Military special forces attract highly fit and intelligent soldiers, and smoke jumping crews attract superbly fit and rugged workers, as do wildland Hotshot and helicopter rappelling crews.

This chapter will help you do the following:

- Determine the physical characteristics of workers.
- Assess the influences these characteristics have on aerobic fitness.
- Discuss why aerobic, muscular, and core fitness training is essential to successful performance in physically demanding occupations.
- Examine demographic trends that influence the pool of qualified workers.

Physical Characteristics of Workers

The American workforce is taller and heavier than ever before. In 1960 men averaged 5 feet 8 inches (173 cm) in height and women averaged 5 feet 3 inches (160 cm); today men and women are 1 1/2 inches (3.8 cm) and 1 inch (2.5 cm) taller, respectively. The average weight for men increased from 166.3 pounds (75.4 kg) in 1960 to 191 pounds (86.6 kg) in 2002. The average weight for women increased from 140.2 pounds (63.6 kg) to 164.3 pounds (74.5 kg) in the same time period. Because we are taller, it makes sense that we weigh more; if the added weight is muscle, it can contribute to work performance. However, the body mass index (BMI), a ratio of weight to height, increased from 25 in 1960 to 28 in 2002. (A BMI above 25 is considered overweight, and a BMI above 30 identifies obesity, unless the waist measurement is less than 40 inches [101 cm]). Recent data classify more than 66 percent of American adults over 20 years of age as overweight or obese, and that is not good for work performance or health (National Center for Health Statistics: www.cdc.gov/nchs/nhanes.htm).

But this book is not about the serious health risks associated with being overweight, including diabetes, hypertension, coronary artery disease, stroke, and even some cancers. This book focuses on the ability to perform physically demanding occupations. This is where aerobic fitness comes into play.

Aerobic Fitness

Aerobic fitness is defined as the ability to take in, transport, and use oxygen. Also called maximal oxygen intake, or $\dot{V}O_2max$, it is usually measured in a treadmill or laboratory bicycle test. The subject performs a progressive test in which work intensity is increased every minute or two by increasing the rate or grade of the treadmill. Expired air is routed to a computerized metabolic measurement system that determines oxygen, carbon dioxide, and expired air volume. The computer corrects air volume for temperature and barometric pressure and calculates oxygen intake each minute. The test is terminated when the oxygen intake plateaus or the subject cannot continue. Aerobic capacity is measured in liters of oxygen consumed per minute (L/min), whereas aerobic power is measured in oxygen consumed per unit of body weight. Aerobic power is calculated by dividing aerobic capacity (L/min) by body weight in kilograms. For example, a 70-kilogram (154 lb) subject achieves a maximal oxygen intake of 3.5 liters per minute: 3.5 / 70 kg = 50 ml · kg^{-1} · min^{-1}, which is that subject's aerobic power.

Aerobic capacity ($\dot{V}O_2max$ in L/min) indicates the maximal size of the aerobic motor. It is related to body size, so larger people, in general, have higher values. Aerobic capacity is related to performance in non-weight-bearing sports, such as rowing and swimming. Aerobic power ($\dot{V}O_2max$

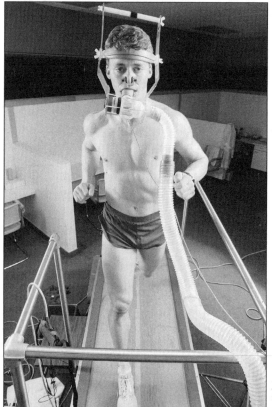

Aerobic fitness is defined as the ability to take in, transport, and use oxygen. It is usually measured in a treadmill or laboratory bicycle test.

©1998 EyeWire, Inc.

in ml · kg⁻¹·· min⁻¹) is related to performance in weight-bearing sports, such as running and cross-country skiing. Aerobic capacity may be important in some occupations, but in weight-bearing tasks, in which workers get around on their feet, aerobic power is often the better indicator of work capacity. In cycling, aerobic capacity predicts performance on a flat course; but in hill climbing, where extra weight is a burden that must be moved uphill, aerobic power has a better relationship to performance (Van Dorn 1987). It indicates the maximal oxygen intake per kilogram of body weight per minute. Aerobic power can contribute to safety. In wildland firefighting, aerobic power is inversely related (r = -0.87) to the time it takes to reach a

safety zone (Ruby et al. 2003). Table 2.1 presents average and standard deviation values for $\dot{V}O_2$max (aerobic power).

When it is used as a standard for physically demanding occupations, the minimum aerobic power score usually ranges from 40 to 50 ml · kg⁻¹ · min⁻¹. Table 2.1 shows variability in scores above the mean; this is where employers will find qualified candidates for physically demanding occupations. You can calculate values below the mean by taking the standard deviation (the difference between the mean and +1 *SD)* and subtracting it from the mean. For example, for 20- to 29-year-old men, the *SD* is 7 so –1 *SD* is 43 – 7 = 36 (see figure 2.1).

Table 2.1　Maximal Oxygen Intake Values (ml · kg⁻¹ · min⁻¹)

	MEN				WOMEN			
Age	Mean	+1*	+2	+3	Mean	+1	+2	+3
20-29	43	50	57	64	36	43	50	57
30-39	42	49	56	63	34	40	46	52
40-49	40	47	54	61	32	38	44	50
50-59	36	43	50	57	29	34	40	46
60-69	33	40	47	54	27	32	37	42

*SD, or standard deviation, is a measure of variability around the mean (average). Plus and minus 1 standard deviation (+/– 1 SD) embraces 68% of the population; +/– 2 SD = 95%; and +/– 3 SD = 99% of the population. Elite male distance athletes score in the high 70s, and elite females approach 70 ml · kg⁻¹ · min⁻¹. This table shows variability above the mean.

Adapted from G. Fletcher et al., 1995, "Exercise standards: A statement for healthcare professionals from the American Heart Association," *Circulation* 91(2): 580-615, G. Duncan, S. Li, and X. Zhou, 2005, "Cardiovascular fitness among U.S. adults: NHANES 1999-2000 and 2001-2002," *Medicine & Science in Sports & Exercise* 37(8): 1324-1328, and S. Gaskill et al., 2001, "Changes in ventilatory threshold with exercise training in a sedentary population: The HERITAGE Family Study," *International Journal of Sports Medicine* 22(8): 586-592.

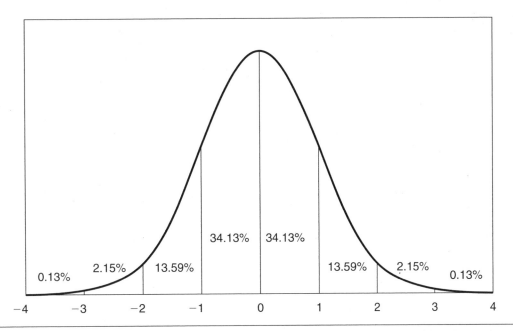

Figure 2.1　The normal (bell-shaped) curve. Percentage values indicate the proportion of distribution in each standard deviation.

Is a high aerobic fitness value the product of heredity or training? How do factors such as gender, age, and body composition influence one's score? And how much can training influence the maximal oxygen intake?

Heredity

Researchers have studied differences in aerobic fitness among fraternal and identical twins and found that intrapair differences were far greater among fraternal than among identical twins. The largest difference between identical twins was smaller than the smallest difference between fraternal twins (Klissouras 1976). Bouchard and colleagues (1999) estimated that heredity accounts for 47 percent or more of the variance in $\dot{V}O_2$max values, and Sundet, Magnus, and Tambs (1994) contended that more than half of the variance in maximal aerobic power is due to genotypic differences, with the remainder accounted for by environmental factors (nutrition and training). This supports the notion that the way to become a world-class endurance athlete is to pick your parents carefully, especially your mother, because maternal transmission accounts for almost 60 percent of the inherited component (Bouchard et al. 1999)!

Picking your parents well may help you achieve elite athlete status, but it is training that dictates how close you get to your potential. We inherit many factors that are important for aerobic fitness, including the maximal capacity of the respiratory and cardiovascular systems, a large heart, a large number of red blood cells and high levels of hemoglobin, and a high percentage of oxidative (slow oxidative and fast oxidative glycolytic) muscle fibers. Mitochondria, the energy-producing units of muscle and other cells, are inherited from the maternal side. Studies indicate that the capacity of muscle to respond to training is also inherited, with improvements in aerobic fitness ranging from 5 to over 30 percent (Bouchard et al. 1988). Other inherited factors such as physique and body composition influence the potential to perform at a high level.

Genes influence potential, but they don't ensure it. The 30,000 genes that form the blueprint of the human body are subject to the influence of the environment and behavior. Your genotype is your genetic constitution, whereas your phenotype is the observable appearance resulting from the interaction of your genotype and the environment. Genetic potential can only be realized when genes are switched on via the process of training. The only way to realize your aerobic potential is to engage in training.

Gender

Before puberty, boys and girls differ little in aerobic fitness, but from then on girls fall behind. Young women average 10 to 20 percent lower in aerobic fitness compared to young men, depending on their level of activity. But highly trained young female endurance athletes are only 10 percent below elite males in $\dot{V}O_2$max and performance times. One reason for the difference between genders may be lower hemoglobin levels (hemoglobin is the oxygen-carrying compound found in red blood cells). Men average about 2 grams more hemoglobin per 100 milliliters of blood (15 versus 13 grams per deciliter [g/dl]), and total hemoglo-

Women are well suited for fat-burning endurance events, and some tolerate heat, cold, pain, and other indignities as well as or better than men do.

CASE STUDY
Fitness Standards for British Columbia Forest Firefighters

The government of the Canadian province of British Columbia established minimum physical fitness standards for its forest firefighters, based on field studies conducted by researchers from the University of Victoria. One was an aerobic standard of 50 ml · kg⁻¹ · min⁻¹. A female previously employed as a firefighter failed to meet the aerobic standard after four attempts and was dismissed. She filed a grievance, and the case went through arbitration and on to the courts. The court of appeals suggested that accommodating women by permitting them to meet a lower standard than men would constitute "reverse discrimination." However, the Supreme Court of Canada disagreed and found that the aerobic standard was discriminatory and that the BC government had not shown that it is reasonably necessary to the accomplishment of the government's general purpose, which is to identify those forest firefighters who are able to work safely and efficiently (*Supreme Court of Canada on appeal from BC Court of Appeals re: BC Government & Service Employees Union v. Government of the province of BC*, 1999).

Evidence accepted by the arbitrator designated to hear the grievance demonstrated that, "owing to physiological differences, most women have a lower aerobic capacity than most men and that, unlike most men, most women cannot increase their aerobic capacity enough with training to meet the aerobic standard." Although this statement is contradicted by numerous studies and women's performance in sport, it was accepted as a statement of fact by the Supreme Court. The data in table 2.1 show that some women are more fit than most men, and with training, many women are able to increase their aerobic capacity to meet the proposed standard. Research on well-trained females provides evidence that with appropriate training, women can achieve high levels of aerobic fitness. Canada is well known for its successful female runners, cross-country skiers, bikers, and triathletes, yet the Supreme Court of Canada chose to ignore the research and the real-world evidence. We will say more about this case, how it was finally resolved, and gender as an issue in physically demanding occupations in chapter 6.

---

bin is correlated to V̇O₂max and endurance. On the other hand, some women have higher values than male endurance athletes have (Sharkey 1984).

Other reasons for their lower level of aerobic fitness may be that women are smaller and have less muscle mass, or that they average more body fat than men (25 percent versus 15 percent for college-age women and men, respectively). Because aerobic fitness is reported per unit of body weight, those with more fat and less lean tissue (muscle) have some disadvantage. Of course, a portion of the difference is sex-specific fat that is essential for reproductive function and health. For these and other reasons (e.g., osteoporosis), women shouldn't try to become too thin. We raise the issue only to explain why the average male has some advantage over the average female in aerobic fitness.

Until the 1970s, women were discouraged, or even banned, from competing in hard work and distance races longer than one-half mile (0.8 km). Officials were afraid that females couldn't stand the strain. Today women run marathons and 100-mile (160 km) races; compete in the Ironman triathlon; and swim, ski, and cycle prodigious distances. Women are well suited for fat-burning endurance events, and some tolerate heat, cold, pain, and other indignities as well as or better than men do. The average man has a small advantage in aerobic fitness, but women can overcome that difference with training.

Age

Figure 2.2 illustrates the decline in aerobic fitness with age. At age 60, V̇O₂max is about 75 percent of the level at age 20. The rate of decline approaches 8 to 10 percent per decade for inactive people, regardless of their initial level of fitness. Those who decide to remain active can cut the decline in half (4 to 5 percent per decade), and those who engage in fitness training can cut that rate in half (2 to 3 percent per decade), at least for a while (Kasch et al. 1988).

A colleague who served as a subject in many studies had a V̇O₂max of 46 ml · kg⁻¹ · min⁻¹ when

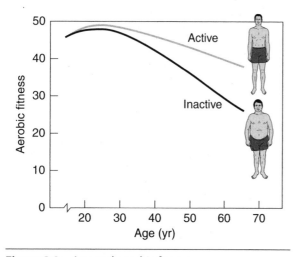

Figure 2.2   Age and aerobic fitness.

Reprinted, by permission, from B.J. Sharkey and S.E. Gaskill, 2007, *Fitness and health*, 6th ed. (Champaign, IL: Human Kinetics), 80; adapted from B.J. Sharkey, 1997, *Fitness and work capacity*, 2nd ed. (Missoula, MT: USDA Forest Service).

he started aerobic training at the age of 30. His score rose to 54 in a few months, then declined slowly for three decades as he continued his active life. When retirement allowed time for even more activity, his fitness was 52, well above his starting point at age 30. Though trainability may decline somewhat with age, exercise gerontologists have shown that fitness can be improved, even after the age of 70 (Kasch, Boyer, Van Camp et al., 1990). And it is never too late to start.

Studies of fire and law enforcement personnel have indicated a poor level of physical fitness in emergency service personnel. Fitness in service personnel declines at the same rate as that of the sedentary population. In one study, body fat levels of emergency service personnel doubled and fitness declined during their 20-year careers (Davis and Starck 1980). With physical decline comes a high rate of heart disease, which, surprisingly, is viewed as a job-related disability. Taxpayers deserve the best public servants available, and passing an entry-level test does not ensure career-long fitness for duty. The solution is to test recruits and follow up with an annual performance evaluation, coupled with a job-related physical maintenance (fitness) program. We'll say much more about age and performance in later chapters.

Body Composition

Ernie, a firefighter, smoked two packs of cigarettes a day, weighed over 250 pounds (113 kg), and bragged about his sedentary lifestyle. When he took a new job-related aerobic fitness test, it was

all he could do to finish with a score in the low 30s. He was disappointed, but he got the message. He stopped smoking, started aerobic training, and paid attention to his diet. Now, two years later, he weighs 170 pounds (77 kg), his fitness score is 58, and he enjoys running road races.

Remember that fitness is calculated per unit of body weight, so if fat increases, aerobic fitness declines. About one-half of the decline in fitness with age can be attributed to an increase in body fat. So the easiest way to maintain or even improve fitness is to get rid of excess fat. For example, if Ernie (at 100 kg [220 lb] and 20 percent fat) loses 10 kilograms (22 lb), or half of his body fat, his aerobic fitness score will go from

$$4 \text{ L} / 100 \text{ kg} = 40 \text{ ml} \cdot \text{kg}^{-1} \cdot \text{min}^{-1}$$

to

$$4 \text{ L} / 90 \text{ kg} = 44.4 \text{ ml} \cdot \text{kg}^{-1} \cdot \text{min}^{-1}$$

Without any exercise, just weight loss, his fitness has improved over 10 percent! Now, if he earns a 25 percent improvement with training, his fitness could rise above 55 ml $\cdot$ kg^{-1} $\cdot$ min^{-1} (44.4 $\cdot$ 25 percent improvement = 11.1 ml $\cdot$ kg^{-1} $\cdot$ min^{-1} + 44.4 = 55.5), an overall increase of 38 percent (55.5 − 40 = 15.5 / 40 = 38.7 percent improvement).

Body Size

Body size is a factor in physically demanding occupations because they often require absolute capacities and performances. Absolute demands, such as being able to lift and handle heavy equipment, place the smaller person, one with a smaller lean body weight, at a disadvantage. The external load of firefighters' personal protective equipment and the energy and muscular demands of the work impose a challenge for smaller employees (see figure 2.3).

Accommodation for smaller people cannot be made without risk to the employee, coworkers, and the public. To perform effectively under these conditions, a smaller person may be required to work at an intensity level higher than that of a larger person, leading to fatigue and injury. Ultimately, the ability of an employee to meet the physical demands of the occupation in a safe, effective, and appropriate manner must be considered (Gaul 2000).

Short stature can increase muscular demands during certain tasks. Figure 2.4 shows two

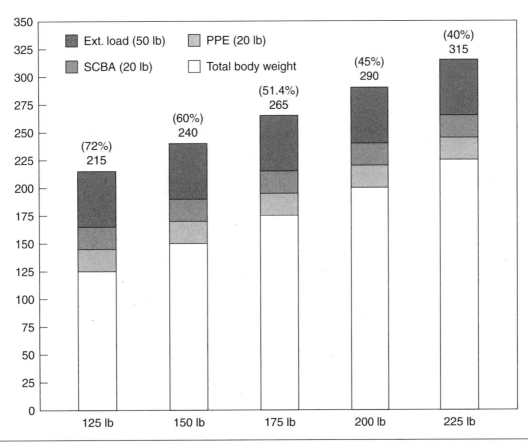

Figure 2.3 Body weight and the impact of external loads. The top numbers in parentheses represent load as a percentage of body weight. The bottom numbers not in parentheses represent total weight (body weight plus load).

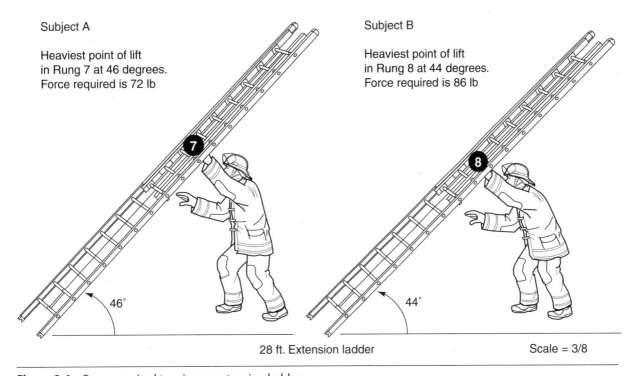

Subject A

Heaviest point of lift in Rung 7 at 46 degrees. Force required is 72 lb

Subject B

Heaviest point of lift in Rung 8 at 44 degrees. Force required is 86 lb

28 ft. Extension ladder

Scale = 3/8

Figure 2.4 Force required to raise an extension ladder.

firefighters raising a 28-foot (8.5 m) extension ladder. The shorter firefighter (subject B), and the one with a lower lean body weight, must exert 20 percent more force to accomplish the task. We will say more about body composition, lean body weight, and body size in later chapters.

Aerobic Training

On average, aerobic training improves aerobic fitness about 25 percent (or more, with weight loss). A person with an aerobic fitness score of 40 ml · kg^{-1} · min^{-1} could raise the value to 50 with 8 to 12 weeks of training. Moderate physical activity leads to improved fitness, enhanced performance, and substantial health benefits. Systematic training leads to a higher level of fitness, performance, and health; and prolonged, systematic training helps people achieve their genetic potential. However, many effects of training can be lost in 12 weeks of detraining, the cessation of systematic training (Coyle, Hemmert, and Coggan 1986). Three weeks of complete bed rest, for example, may cause a fitness decline of 29 percent, or almost 10 percent per week, but the loss can easily be restored with a return to training. A year-round program is the ideal way to maintain fitness, performance, and health.

The capacity to improve aerobic fitness is not limited by age or gender. At 81 years of age, Eula Weaver had a heart attack to add to her problems of congestive heart failure and poor circulation. Unable to walk even 100 feet (30.5 m) at first, she worked up to jogging a mile (1.6 km) each day and

riding her stationary bicycle several more miles. She even lifted weights several days a week. At 85 she won the gold medal for the mile run in her age group at the 1974 Senior Olympics.

We will say much more about fitness and work capacity in chapter 11, Physiology of Work. Now let's examine the factors that influence the components of muscular fitness.

Muscular Fitness

In the early 1970s, few, if any, women engaged in wildland firefighting, so researchers developing a work capacity test saw no need to include a measure of muscular fitness. Instead they focused on aerobic power as the factor most likely to limit performance in this arduous, long-duration job. However, when the aerobic test was implemented in 1975, it coincided with the entry of women into the workforce. Soon thereafter, crew bosses became concerned that some firefighters, many of whom were women, lacked sufficient strength or muscular endurance for the job. A large field study analyzed the relationship between muscular fitness and performance and found that measures of strength and muscular endurance, as well as lean body weight, were significant predictors of work performance (Sharkey 1981).

Strength

Strength is defined as the maximal force that can be exerted in a single voluntary contraction. The force depends on a number of factors, such as muscle mass or the cross-sectional area of muscle; the number of contracting muscle fibers and their contractile state (length or fatigue); and the mechanical advantage of the bony lever system. Most of these are easy to explain. More muscle fibers equals more girth and more force. A stretched muscle exerts more force than a rested muscle (probably because of elastic recoil and a favorable alignment of contractile proteins); a rested muscle exerts more force than a fatigued muscle; and mechanical factors such as limb length conspire to magnify force or speed. Several other factors, including gender, age, fiber type, and training deserve mention.

Gender

Until 12 to 14 years of age, boys are not much stronger than girls. Thereafter, the average male gains an advantage that persists throughout life. Is the difference due to the increase in the male hormone testosterone at puberty? Perhaps. The average male has 10 times the testosterone found in the average female. Testosterone is an anabolic (growth-inducing) steroid that helps muscles get larger. Studies show that untrained women have approximately half of the upper-body strength and approximately two-thirds of the lower-body strength of men (Miller et al. 1993).

Young women average twice the percentage of fat as men (25 to 28 percent versus 12.5 to 15 percent). When you look at strength per unit of lean body weight (body weight minus fat weight), women have slightly stronger legs, but weaker arms (30 percent below men's values). Wilmore (1983) suggested that because women use their legs as men do (walking, climbing stairs, bicycling), their leg muscles are similar in strength. Because fewer women use their arms in heavy work or sport, their strength lags behind in this area. As more women engage in upper-body strength training for sport or occupational purposes (police work, firefighting, or construction), their strength will certainly come closer to that of men. In fact, recent research with cross-country skiers showed that when elite female athletes were evaluated for strength per unit of lean body weight, they had 97 percent of the strength of male athletes. When women train their upper bodies, they become more comparable with men (Steve Gaskill, personal communication).

Age

Without a systematic program of resistance training, strength reaches a peak in the early 20s and declines slowly until about age 60. Thereafter, the rate of decline usually accelerates, but it doesn't have to. When muscles are trained systematically throughout one's life, strength hardly declines at all, even into the 60s. Champion weightlifters have achieved personal records in their 40s. Auto mechanics retain grip strength into their 60s. Training before puberty leads to improvements that are mostly due to changes in the nervous system (neurogenic factors include reduced inhibitions and learning how to exert force). Training after puberty combines nervous system changes with changes in the muscle tissue (myogenic changes). Because testosterone levels decline in old age, many physiologists thought that senior citizens would be limited to neurogenic changes when they engaged in strength training. However, a study of very elderly people (72 to 98 years) has shown that resistance training leads to increased strength, muscle mass, and mobility (Fiatarone et al. 1994). Strength training results in increases in protein synthesis, strength, and muscle size, and improved performance.

Muscle Fiber Types

Human muscle has two types of muscle fibers, slow-twitch and fast-twitch (see figure 2.5). The larger, faster-contracting fast-twitch fibers have a greater potential for the rapid development of tension. People with a higher percentage of fast-twitch fibers have a greater potential for force development. Studies of human muscle tissue revealed that powerlifters have twice the area of fast-twitch fibers as nonlifters. The size can be attributed to heredity and to training. The effect of strength training on muscle fiber types has yet to be completely understood; current evidence indicates that both types grow larger with training, but growth of the fast fibers is more pronounced. Strength training improves the capabilities of both types, but doesn't seem to change one type into another. More information about muscle fibers can be found in chapter 11, Physiology of Work.

Strength Training

Progressive resistance training, which leads to improvement in strength, typically involves using a weight that can be lifted 8 to 12 times until the point of muscular fatigue. This kind of training leads to increases in contractile proteins (actin

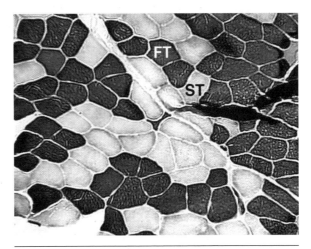

Figure 2.5 Fast- and slow-twitch muscle fibers. FT = fast-twitch; ST = slow-twitch.

Reprinted, by permission, from B.J. Sharkey and S.E. Gaskill, 2006, *Sport physiology for coaches* (Champaign, IL: Human Kinetics), 47.

and myosin) and tougher connective tissue. A previously sedentary person on an adequate diet (energy, protein) can increase strength 50 percent or more in six months. Concentrated training can lead to similar gains in less time. The strength gains are specific to the training; they take place only in the muscles trained. For this reason, strength training should involve muscle groups used on the job. See *Sport Physiology for Coaches* (Sharkey and Gaskill 2006) for a comprehensive guide to strength training.

A recent study compared strength and muscle size gains in 585 men and women after 12 weeks of training. There was considerable variability in the responses: Size changes ranged from –2 to 59 percent, and strength gains ranged from 0 to 250 percent. Men experienced 2.5 percent greater gains for cross-sectional areas of muscle, as measured by magnetic resonance imaging, and greater absolute strength gains when compared with women. Women did better than men in relative gains in strength (percentage gains). There was a wide range of response to resistance training, with some subjects doubling their strength (Hubal et al. 2005). Heredity plays a role in the response to training,

as does one's initial muscular fitness level. Those with less strength at the start of a training program make greater percentage gains.

If strength is required on the job, it makes sense to engage in job-related strength training. When strength becomes adequate for the task, it may be time to switch to muscular endurance training.

Muscular Endurance

Muscular endurance implies the ability to persist. It is measured in repetitions of submaximal contractions. Muscular endurance is essential for success in many work activities. Once a person has sufficient strength to perform a repetitive task, additional improvement in performance will depend on muscular endurance, or the ability to persist. As mentioned earlier, fast-twitch muscle fibers are strong, but they fatigue more readily than slow-twitch fibers do. Endurance training improves the ability of slow- and fast-twitch fibers to sustain contractions. Muscular endurance resides in metabolic adaptations, the ability to utilize oxygen, and learned neuromuscular efficiency (Sharkey and Gaskill 2007).

Endurance is gained with less resistance (under 66 percent of maximal force) and more repetitions (15 to 25 repetitions maximum, or RM). Muscular endurance is easily improved with training. Although it is difficult to double the number of pull-ups (that takes strength), it is easy to double the number of push-ups (they require endurance). Improved strength and endurance make work seem easier.

Because different occupations use different muscle groups in different ways, there is no single, agreed-upon list of strength and endurance tests. Table 2.2 presents some strength, muscular endurance, flexibility, and power tests. We provide job-related muscular fitness test examples in future chapters and training suggestions in chapter 12.

Core Training

Core stability is related to the strength, muscular endurance, and muscle balance of the core muscles. The core muscles control the trunk, pelvis, and shoulders, stabilizing the body and providing a foundation for forceful movements by the arms and legs. Core stability receives considerable attention in the training of athletes and should receive more in the workplace. Core training can

IIIIIIIIIII JOB-RELATED IIIIIIIIIII TRAINING FOR WOMEN

Most women and men have the ability to engage in physically demanding occupations if they are willing to train. When U.S. Army researchers recruited 41 female soldiers to volunteer for a 24-week training program consisting of weightlifting, 5-mile (8 km) backpack hikes, 2-mile (3.2 km) runs, and additional drills, pre- and posttraining tests showed significant improvements in lifting (40 lb [18 kg] box to 52 in. [132 cm], height of truck bed), load carriage (carry 40 lb [18 kg] 25 feet [7.6 m] and lift to 52 in. [132 cm] height), vertical jump (power), and aerobic fitness. Also notable was their 32.5 percent improvement in 2-mile (3.2 km), 75-pound (34 kg) backpack speed, to 4.4 miles per hour (7 km/h), well above the 2.5 miles per hour (4 km/h) rate listed in the *Army Field Manual*. Before training the volunteers could lift 70 percent of what male soldiers could in a maximal lift to 52 inches (132 cm). After 24 weeks of training, the female soldiers lifted 91 percent of male values (Harman et al. 1997).

Table 2.2   Muscular Fitness Tests

Test	MEN			WOMEN		
	Low	Medium	High	Low	Medium	High
Upper-body strength: Pull-up*	<6	6-10	>10	<20	20-30	>30
Endurance: Push-up	<20	20-40	>40	<10	10-20	>20
Trunk endurance: Sit-up	<30	30-50	>50	<25	25-40	>40
Leg strength: Leg press	<400	400-550	>550	<300	300-450	>450
Power: Vertical jump	<17	17-23	>23	<12	12-17	>17
Flexibility: Sit and reach toes	No	Yes	Beyond	No	Yes	Beyond

*Women do modified pull-ups.

Data from University of Montana Human Performance Laboratory.

Core training can improve a worker's performance and reduce a worker's risk of back and other musculoskeletal problems.

improve worker performance and reduce the risk of back and other musculoskeletal problems. It provides a strong base from which to develop forceful movements, transferring power from the central core. Workers should train their abdominal, back, shoulder, and hip muscles to strengthen their core, and then maintain a year-round job-related maintenance program (Sharkey and Gaskill 2006). Core training is essential in sport and in physically demanding occupations.

Demographic Trends

A number of demographic trends influence the availability of qualified workers for physically demanding occupations. They range from the obvious, including lack of physical activity, overweight, obesity, and age, to the less obvious, including a decline in compensation and the well-meant activities of advocacy groups and the courts.

Inactivity and Overweight

There is no question that the American adult population has been gaining weight for the last 15 to 20 years. Recent data reported by the U.S. National Center for Health Statistics identified 66.5 percent of the adult population (over 20 years of age) as overweight or obese, including 71 percent of men and 62 percent of women (www.cdc.gov/nchs/nhanes.htm). The men's numbers rose 4 percent in five years, whereas the women's numbers held steady. Obesity (BMI of 30 and above) increased 3.6 percent in men to 31.1 percent, whereas women held steady at 33.2 percent. The main reason for the rise in overweight is an imbalance between energy intake (food) and energy expenditure (physical activity). Inactivity and overweight both diminish the numbers of qualified workers available for physically demanding jobs and increase health care costs.

Those who are inactive fail to develop the energy and muscular fitness required for hard work. And few workers of any age seem willing to engage in the training it takes to develop these capacities. Employers and the military have found that many young recruits are not familiar with hard work. Young adults have grown up with television, computers, and video games. Few have had jobs, let alone hard ones. Add overweight or obesity to the picture, and it becomes even more difficult to train for or perform physically demanding tasks.

IIIIIIIIIIIII **BATTLE OF THE** IIIIIIIIIIIII
BULGE

Even the U.S. military is participating in the trend toward overweight. Over a three-year period, the percentage of military classified as overweight (BMI of 25 or above) rose from 50 to 54 percent, reflecting the trend in the general population. This increase occurred in spite of an increase in physical activity, suggesting that the personnel were eating more. Overweight military personnel in this large representative sample were more likely to be male, older, African American or Hispanic, married, and enlisted personnel (Lindquist and Bray 2001).

Aging

The United States is an aging population. So-called baby boomers are approaching retirement age, and the over-65 group is the fastest-growing segment of the population. As older workers retire from fire-fighting and law enforcement positions, the agencies turn to the less active, overweight segment just discussed. Some states recruit young adults into summer training camps to teach skills and work habits. California has a fire training academy that teaches firefighting skills and prepares participants for the rigors of the fire line.

Compensation

At the beginning of this chapter we said that the physically demanding jobs will attract workers so long as the jobs are rewarding and the compensation is adequate for the effort. Some forms of hard work pay low wages and have thus lost their appeal to most U.S. citizens. The pay is low because employers are able to hire undocumented (illegal) workers willing to work for lower wages. These employers claim that Americans are not interested in the work, and yet American workers are available for arduous firefighting and mining jobs, jobs that pay a fair wage. Farm work, meat cutting, hotel work, landscaping, and a growing number of construction jobs are often done by illegal immigrants, causing a downward pressure on compensation. This trend is sustained by porous borders; employers who avoid paying fair compensation, including health and retirement benefits; and politicians who tolerate illegal workers while they collect campaign contributions from wealthy employers.

As young men, the authors of this book were glad to do farm work, construction, and meat cutting so long as the compensation was adequate. Today, with better compensation, U.S. employers could provide meaningful employment for citizens, and perhaps we could begin to coax American teenagers away from video games into the meaningful world of work. Teen employment could reduce our need for immigrant labor and begin to reduce the incidence of overweight and obesity. In the words of Theodore Roosevelt,

> In the last analysis, a healthy state can exist only when the men and women who make it up lead clean, vigorous, healthy lives; when the children are so trained that they shall

endeavor, not to shirk difficulties, but to overcome them; not to seek ease, but to know how to wrest triumph from foil and risk.

Theodore Roosevelt in "The Strenuous Life"

Advocacy Groups and the Courts

Advocacy groups and the courts are making it more difficult to maintain performance standards in physically demanding occupations. San Francisco developed a physical capacity test for prospective firefighters. When few women were able to pass the test, it was dropped and a new test was developed. The new test was so easy that some junior high school students could pass it. The city could have provided pretest training for recruits or given them more time to meet legitimate standards, but they did not. Instead it lowered the standards and accepted less capable recruits. We deal with legal aspects of the employment process in chapter 3, and other legal issues in chapter 16.

Summary

This chapter determined the physical characteristics of workers, including their aerobic and muscular fitness, body composition, and lean body weight. It assessed the influences of age and gender on work capacity, noting that training can retard the effects of aging and minimize the gender differential.

Consider the effects of age on a marginal firefighter recruit. In a few years, without physical training, he will fall below minimal aerobic and muscular fitness standards. The employee will regress in fitness and health and gain body fat at the same rate as those in the sedentary population. He will become a candidate for injury, spend more time on sick leave, and raise the cost of workers' compensation. And he is likely to suffer overweight, hypertension, diabetes, and heart disease, leaving the job early on a disability or dying on the job. If he smokes, he may get lung disease and blame it on the job. This is not a good investment for taxpayers. Job-related fitness training is essential in a physically demanding occupation. It improves fitness, performance, health, and morale.

Demographic trends have begun to influence the pool of qualified workers, including lack of activity, overweight, obesity, aging, and of course, diminishing compensation. Chapter 3 considers some of the laws and regulations that govern the employment of workers in physically demanding occupations.

Chapter 3

Employment Opportunity

© AP Photo/Mike Derer

hroughout the history of the United States, a number of laws and executive orders have addressed employment, including the Civil Rights Acts of 1866 and 1871 and the Civil Service Act of 1883, which established merit and fitness as the only factors that would determine entry and advancement within the federal government. An executive order by President Roosevelt in 1940 prohibited discrimination in the federal government, and subsequent orders in 1941 prohibited discrimination in defense and procurement contracts. Presidents Truman, Eisenhower, and Kennedy issued orders that further focused on employment practices. Since 1963 the number of laws and regulations governing the employment process has expanded dramatically. The laws cover all employment, but they have significant implications when hiring personnel for physically demanding occupations.

In this chapter we consider important federal laws and regulations and how they have been applied in the workplace. The purpose of these employment laws and regulations is to prohibit illegal discrimination in employment and provide equal employment opportunities for all citizens. Illegal or unfair discrimination occurs when employment decisions are based on race, ethnicity, religion, age, sex, or disability rather than on knowledge, skills, and abilities. Because some states, counties, and cities have their own employment laws, it is important to be familiar with state and local laws on employment and testing.

Finally, we introduce the U.S. government's Uniform Guidelines on Employee Selection Procedures (UGESP), which provide a framework for the development of employee selection assessments.

This chapter will help you do the following:

- Identify the key employment laws that govern the employment process.
- Explore the concepts of discrimination and adverse impact in the hiring process.

Employment Laws

In the United States, the Equal Employment Opportunity Commission (EEOC) is responsible for enforcing federal laws prohibiting employment discrimination. Other federal laws that are not enforced by EEOC also prohibit discrimination against federal employees. In this section we review key laws and show how they might influence employment in physically demanding occupations. We use italics to identify important words in the laws that relate to hard work.

Equal Pay Act (EPA) of 1963

This act protects men and women who perform substantially equal work in the same organization from sex-based wage discrimination (www.eeoc.gov/policy/epa.html).

Title VII of the Civil Rights Act (CRA) of 1964

Amended in 1972, Title VII prohibits employment discrimination based on race, color, religion, sex, or national origin. This act affects employers with 15 or more employees, employment agencies, and labor unions. The 1972 amendment indicates that workplace tests can be used to make employment

Lance Cpl. Phillip M. Cox, administrative specialist, Base Inspector's Office, hopes one day to become the first African American commandant of the U.S. Marine Corps.

decisions, but the tests must not discriminate against protected groups (www.eeoc.gov/doc.php).

Age Discrimination in Employment Act (ADEA) of 1967

The ADEA prohibits discrimination against employees or applicants who are 40 years of age or older in all aspects of employment, except when age is a bona fide occupational qualification (BFOQ). The ADEA affects employers with 20 or more employees, employment agencies, and labor unions. Discrimination in testing is prohibited. If a worker charges discrimination, the employer may defend the practice if the job requirement is deemed a business necessity. In spite of a comprehensive study that concluded that age was not a BFOQ in public service occupations (Landy 1992), Congress has exempted law enforcement personnel and firefighters from the law, along with military personnel (www.eeoc.gov/types/age.html).

Rehabilitation Act of 1973

Sections 501 and 505 of the Rehabilitation Act prohibit discrimination against qualified people with disabilities who are employed by the federal government (www.eeoc.gov/policy/rehab.html).

Civil Service Reform Act (CSRA) of 1978

The Civil Service Reform Act contains a number of prohibited personnel practices that were designed to promote fairness in federal personnel practices and prohibit discriminating against employees on the basis of race, color, national origin, religion, sex, age, or disability. The act also provides that personnel practices not be based on conduct that does not adversely affect job performance, such as marital status, political affiliation, and sexual orientation (www.dol.gov/regs/compliance/olms/complcsra.htm).

Americans with Disabilities Act (ADA) of 1990

The ADA prohibits employment discrimination against qualified people with disabilities in the private sector and in state and local government. The law affects employers with 15 or more employees, employment agencies, and labor unions. Discrimination includes failure to provide reasonable accommodation to people with disabilities when doing so would not pose undue hardship. A qualified person with a disability is one who can perform the essential functions of the job, with or without reasonable accommodation. *Disability* is broadly defined to include physical or mental impairment that substantially limits one or more major life activities (e.g., seeing, hearing, or talking).

This comprehensive act, which also covers accessibility, expressly prohibits medical inquiries or a medical examination prior to making a conditional job offer. After a job offer, a medical examination is allowed if it is work related and justified by business necessity. A physical performance test, in which the applicant demonstrates

CASE STUDY

Violating the Americans with Disabilities Act

A nationwide package delivery company decided to evaluate recruits on the basis of a job-related physical performance test. Low back and other musculoskeletal problems were causing excess sick leave and workers' compensation costs. The test determined the applicant's ability to move 50-pound (22.7 kg) boxes from a 36-inch-high (91.4 cm) conveyor belt to a 52-inch (132 cm) truck bed. The score was the number of boxes moved in five minutes. Recruits completed a one-page health risk assessment prior to the test to screen for serious health problems. Because no medical exam was to be given, the company decided to check evaluees' blood pressure before the test and their heart rate immediately thereafter. The heart rate information would tell them who was working close to their physical limitations. When the company attorney evaluated the proposed test, she decided that the procedure, as outlined, violated the Americans with Disabilities Act (ADA).

his or her ability to perform actual or simulated job tasks, is not considered a medical exam and can be administered at the preoffer stage. The test should not use information regarding the person's health. Physiological responses to performance (e.g., blood pressure or heart rate) would constitute a medical examination and would not be allowed prior to an offer of employment (www.eeoc.gov/policy/ada.html).

Title I of the Civil Rights Act (CRA) of 1991

Title I reaffirms and extends the CRA of 1964. It requires evidence of the job-relatedness and business necessity of test procedures that cause adverse or disparate impact. Title I prohibits different cutoff scores for different groups and other types of score adjustments, and makes compensatory and punitive damages available as a remedy for intentional discrimination (www.eeoc.gov/policy/vii.html).

IIIIIIIIIIIIIII **DAVIS' 10** IIIIIIIIIIIIIIIIIIIII **EMPLOYMENT TRUTHS THAT ARE AXIOMATIC WITHIN LAW ENFORCEMENT**

1. Employment in law enforcement is not a right; it is a privilege.
2. Although largely sedentary, the job does have, at times, profoundly demanding physical requirements.
3. The physical capabilities of the criminal element define the activities and the response of the law enforcement officer.
4. The physically fit officer has a greater probability of success than his or her less fit counterpart (i.e., more is better).
5. Physical fitness is an attribute that, if not maintained, degrades over time.
6. There are no prohibitions against applicants or incumbents improving their physical fitness commensurate with the requirements of the job.
7. No physical ability test is ever perfect.
8. There is no such thing as a physical ability test that does not have adverse impact.
9. The fact that a test has an adverse impact does not make it illegal.
10. There can be no physical fitness without a physical training program.

For a comprehensive examination of these 10 employment axioms, see appendix A.

Uniform Guidelines on Employee Selection Procedures

The employment laws we've just discussed refer to qualified people, those who are able to perform the essential functions of the job. To identify qualified people, several laws refer to the performance of workplace or simulated job tasks. Because of the energy and muscular demands and risks associated with physically demanding occupations, it is important to consider the physical capabilities of applicants. To guide the development of employee selection procedures, the U.S. federal government issued the Uniform Guidelines on Employee Selection Procedures (UGESP) (1978).

In 1978 the Equal Employment Opportunity Commission, the Labor Department, the Justice Department, and the Civil Service Commission (now the Office of Personnel Management) issued the federal Uniform Guidelines. The guidelines incorporate a single set of principles that are designed to assist employers, labor organizations, employment agencies, and licensing and certification boards in complying with federal laws prohibiting employment practices that discriminate on the grounds of race, color, religion, sex, and national origin. The guidelines are legally binding under a number of laws, including Executive Order 11246, and the courts, the EEOC, and federal and state agencies apply the guidelines when enforcing these laws (Uniform Guidelines on Employee Selection Procedures 1978).

The guidelines cover employers with 15 or more employees, labor unions, and employment agencies, as well as contractors and subcontractors to the federal government and organizations receiving federal aid. They apply to tests and procedures used to make employment decisions, including hiring, promoting, and termination, as well as some training. The Uniform Guidelines have significant implications for personnel assessment and job-related testing. In this section we consider some of the provisions of the guidelines.

Discrimination

The use of any selection procedure that has an adverse impact on the hiring, promotion, or employment of members of any race, sex, or ethnic group is considered to be discriminatory and inconsistent with the guidelines, unless the procedure has been validated or modified in accordance with the guidelines.

It should be noted that the term *discriminate* has several meanings: One is "to make a distinction in favor of or against a person or thing on the basis of prejudice." Another meaning of the term is "to distinguish accurately" (*Webster's Desk Dictionary* 1983). In that context, a test must discriminate if it is to serve in the selection process. If all students get 100 percent or zero on a test, the test doesn't discriminate. Similarly, if all job applicants easily pass or all fail a job-related test, it doesn't distinguish or differentiate. A test for employment in a physically demanding occupation should discriminate

THE AA/EEO PLAN
The Constitution guarantees equal opportunity, not results.
—Anonymous

Many public and private agencies and institutions have adopted affirmative action/equal employment opportunity (AA/EEO) plans to provide for and promote equal employment opportunity in compensation and conditions of employment, without discrimination based on race, creed, color, national origin, gender, age, disability, or marital status. Some also include sexual orientation, genetic predisposition, and Vietnam era veterans, as well as a commitment to diversity. The policy applies to all employment practices and actions, including recruitment, job application, examination and testing, hiring, training, transfer, reassignment, and promotions. An AA/EEO plan usually sets forth programs and goals for increasing the representation of historically excluded groups.

Departments or agencies with AA/EEO plans should follow them. Such plans often specify recruitment strategies designed to elicit responses from applicants in protected classes. Employers should let applicants know whether they provide pretest training opportunities and the available times and locations.

among those who can and cannot do the work, regardless of race, color, national origin, religion, sex, age, or disability.

Adverse Impact

Adverse, or disparate, impact is defined in the Uniform Guidelines as a selection rate for any race, sex, or ethnic group that is less than four-fifths (4/5) or 80 percent of the rate for the group with the highest selection rate. If the pass rate for men on a work performance test is 90 percent, the test will have an adverse impact if women's pass rate falls below 72 percent (90 × 80 percent = 72 percent). When the pass rate is above 80 percent, adverse impact doesn't exist and the guidelines do not require that the job-relatedness of the test procedures be demonstrated. However, when adverse impact does exist, the guidelines recommend that the test be

modified, unless it is clearly justified by business necessity. Chapters 4 and 5 delineate steps in the development and validation of job-related tests. A well-developed test differentiates among levels of performance but does not discriminate on the basis of race, color, national origin, religion, sex, age, or disability.

Adverse impact, as defined in the guidelines (80 percent), is an arbitrary standard that should be subject to interpretation. An assessment procedure that causes adverse impact may be used if there is evidence that the test is job related and its use is justified by business necessity. For example, if a class of firefighter recruits has difficulty with a test because they lack sufficient muscular fitness to work while wearing almost 50 pounds (22.7 kg) of equipment, and the equipment is essential to job performance and safety (business necessity), the test can be used. In chapter 2 we described gender differences in upper-body strength. Because the average woman has approximately 60 percent of the upper-body strength of the average man, the 80 percent cutoff could cause adverse impact. Work tasks that are predominately aerobic are less likely to cause adverse impact.

The business necessity of a test is established by showing that the test is essential to the safe and efficient operation of the business and that there are no alternative procedures available that have sufficient validity to achieve business objectives, but with less adverse impact. To establish job-relatedness, validity, and business necessity, employers must maintain the documentation described in the guidelines, including the steps in test development, the validation approach, alternative procedures investigated, the numbers tested in each protected group, adverse impact, and more.

Finally, consider the Equal Pay Act of 1963, which protects men and women who perform substantially equal work in the same organization from sex-based wage discrimination. Studies of physically demanding occupations indicate that some workers can perform work tasks several times faster than others can. In a study of wildland firefighters, the most fit and best trained employees performed the tasks three to four times faster than less fit and less skilled workers did. If workers perform similar tasks at significantly different rates, does that constitute substantially equal work and fall under the terms of the act? If a fire were approaching your house and you could select the firefighters, which would you choose: the elite crew

A female wildland firefighter. The Equal Pay Act of 1963 protects men and women who perform substantially equal work in the same organization from sex-based wage discrimination.

that can knock down the fire in 15 minutes or the marginal crew that could take an hour or more? It's your house; you decide.

Summary

This chapter outlined the legal context governing the employment process; discussed how race, color, national origin, religion, sex, and age affect the process; and indicated how the law is intended to protect qualified people with disabilities.

We introduced the federal Uniform Guidelines on Employee Selection Procedures, which were designed to help employers and others comply with requirements of federal law prohibiting employment practices that discriminate. In a discussion of the term *discrimination*, we pointed out that the word has several meanings. The laws prohibit employment discrimination based on race, color, religion, sex, age, national origin, and disability. Tests, however, must discriminate—that is, distinguish or differentiate among applicants—if they are to serve a useful purpose.

The concept of adverse impact was described with the 80 percent (four-fifths) rule, a selection rate for any race, sex, or ethnic group that is less than four-fifths (4/5), or 80 percent, of the rate for the group with the highest selection rate. If a

CASE STUDY

Criterion Task Test

A human factors organization was tasked with developing job-related physical performance standards for the entry-level position of firefighter for the City of St. Paul, Minnesota. A comprehensive job task analysis (JTA) was conducted for the position of firefighter using a department-wide survey. The 184 incumbent respondents identified critical, arduous, and frequently performed job tasks that provided the basis for the development of the criterion task test (CTT), a five-item sample of frequently performed or critical fireground evolutions (hose carry, fan carry, forcible entry, hose advance, and victim rescue).

A group of 25 subjects, 16 men from the St. Paul Fire Department and 9 women from the Phoenix Fire Department, were tested on an array of physiological performance indicators, including measures of aerobic power, muscular strength, and body com-

position prior to performing the CTT. The cohort of women professional firefighters from the Phoenix Fire Department performed the identical protocols, and their data were merged with those of Saint Paul for the purposes of statistical analysis. The energy and metabolic systems required to perform the CTT while wearing turnout gear (helmet, coat, pants, gloves, boots) and self-contained breathing apparatuses (SCBAs) were measured through the collection of expired air samples and blood lactate levels as the subjects performed the test course. Additionally, the group completed a posttest questionnaire that included questions about the relative arduousness, perceived exertion, and realism of the CTT.

The pace for the performance of the CTT was derived by having 10 supervisory personnel (district and deputy chief officers) independently view videotapes displaying a range of nine different paces on three different videotapes and indicating acceptable levels of performance. The establishment of time limits to complete the CTT for selection purposes was driven by the following factors:

- The fact that firefighter positions should be filled by the most qualified applicants
- Job-related design of the CTT
- Actual energy cost of the tasks and physiological requirements during CTT performance
- Limitations imposed by turnout gear and self-contained breathing apparatuses (SCBAs)
- Margin between physical performance requirements and physical capacity to avoid injury and allow emergency response
- Performance expectations driven by public safety mandates
- Reasonable accommodation for protected classes

The people with the fastest times on the CTT had the lowest lactate levels, whereas those who took longer than six minutes had higher lactate levels. The data also revealed that the greater the person's aerobic capacity was, the lower his or her lactate level was at the completion of the CTT. In other words, the best performers were the most fit, had the highest energy contributions from the aerobic system, and were most capable of clearing lactate. Above the threshold of six minutes, the higher lactate levels suggest a lower aerobic capacity or smaller muscle mass and less ability to recover between evolutions. These findings support the validity of using an upper limit of time (six minutes) as a criterion for selection.

The recommended scoring method for the physical performance test is as follows:

Under 5 minutes	Highly qualified
5 to 6 minutes	Qualified
Over 6 minutes	Unqualified

It was recommended that a 1.5-mile (2.4 km) run be used as an initial screen to prequalify candidates before testing them on the CTT. A cutoff of 11 minutes and 40 seconds, corresponding to a maximum oxygen uptake of 45 milliliters per kilogram per minute, should be used to qualify for the CTT test. The score of 45 has been reported as the level of aerobic fitness required for firefighting. This prescreen qualification serves two purposes: It assesses the requisite level of aerobic fitness, and it limits the list of candidates tested on the CTT. It was also recommended that this test instrument (CTT) be used on an annual basis as a condition of continued employment. (Contributed by Bill Gilman, PhD, City of St. Paul Fire Department)

CASE STUDY

Physiological Responses

A related study was conducted to examine the physiological responses of firefighters during a simulated rescue of hospital patients and to relate the firefighter's performance to their endurance, strength, and working technique (vonHeimburg, Rasmussen, and Medbo 2006). Each firefighter ascended six floors wearing personal protective clothing and SCBA and carrying tools (an extra load of 81 lb [37 kg]). They then rescued six people by dragging them across the floor. Oxygen intake and heart rate were recorded throughout the tests, and blood lactate was measured during and just after the rescue. A peak blood lactate level of 13 millimoles indicated that the five- to nine-minute rescue was an all-out effort. The peak oxygen intake of 3.7 liters per minute was 84 percent of the $\dot{V}O_2$max. Large and heavy firefighters carried out the task faster than smaller ones did. The $\dot{V}O_2$max and the dragging technique were both related to the rescue performance. A minimum aerobic capacity of 4.0 liters per minute was recommended for firefighters.

(Subjects in this Norwegian study averaged 4.4 L/min for $\dot{V}O_2$max. A 176-pound, 80-kilogram, firefighter would need an aerobic power score of 50 ml · kg^{-1} · min^{-1} to achieve an aerobic capacity of 4.0 liters per minute [80 kg $\times$ 50 ml · kg^{-1} · min^{-1} = 4.0 L/min]. A 154-pound, 70-kilogram, firefighter would need a $\dot{V}O_2$max of 57 ml · kg^{-1} · min^{-1}, a value well above the average for young men or women in the United States.)

test has adverse impact, it should be eliminated, modified, or justified. The test can be used if the procedure is job related and valid for selecting better workers, and there is no equally effective test that has less impact.

The next part of this book addresses the development, validation, and implementation of job-related tests for physically demanding occupations.

PART II

Test Development and Validation

© David Sanders

That a man must be physically sound for his work we know,
but a standard of soundness has never been defined . . .
it is urgent that a simple but effective method be used
by all employing officers to ensure the rejection of the clearly unfit.

Coert Dubois, 1914, U.S. Forest Service

Employers often go to considerable expense and effort to select new employees. They use written tests, psychological tests, drug tests, and, after conditional job offers, medical examinations. For physically demanding occupations, many employers use job-related tests that assess prospective employees' capacity to perform the work. Once employees are hired and trained, they begin a trial, or probationary, period, during which employers determine whether they are suited for the job. In many occupations, there is no further evaluation of the physical capacity to perform hard work throughout the worker's career.

The chapters in part II outline the value of initial and periodic evaluations, the test development process, the complexities of test validation, and issues related to testing. We consider research designs, subject matter experts, job task analysis, essential functions, confidentiality, and management strategies. We compare the effects of selecting the best applicants to accepting those who pass a test of minimum qualifications. And we document the value of annual or periodic evaluations.

In this part we also address the ways the test development process is consistent with existing employment laws. We indicate how establishment of job-related standards contributes to performance, and how work capacity responds to training or to inactivity. We also discuss instances in which the courts have ignored human physiology and made decisions that are invalid and discriminatory. But courts are not the only ones to make mistakes. Labor unions have negotiated contracts that effectively prohibit annual tests of work capacity or mandatory involvement in job-related fitness programs. This is especially true in law enforcement and firefighting. When this occurs, those who pay the salaries—taxpayers—cannot be certain that the employees are fit to perform the critical tasks.

Chapter 4
Job-Related Tests

© AP Photo/Anderson Independent-Mail, Kendra Waycuilis

In the world of sport, the coach seeks recruits with the physical skills and abilities that are associated with success. Tryouts are held, and those who show promise are invited to join the team. Those lacking sufficient capability are cut from the squad, regardless of race, color, religion, or national origin. In this realm of physical meritocracy, coaches or teams are not sued for discrimination. But in the world of hard work, grievances may be filed and employers may be sued. For these and other reasons, it is important to understand how to evaluate and select prospective workers for physically demanding occupations.

This chapter will help you do the following:

- Appreciate the reasons for using job-related tests.
- Identify legal concerns such as liability and negligence.
- Understand the test development process including the job task analysis and designing and evaluating job-related tests.

Why Test?

Employee salaries and associated overhead costs (health insurance, retirement) are major expenses in most organizations. It makes sense to hire the most capable employees, especially in physically demanding occupations. Administering job-related tests is an effective method for ensuring that the most capable employees are hired. Reasons to test include ensuring the safety and health of employees and maintaining productivity.

Safety

Work duties pose a unique combination of physical demands and risks. Although in the United States the Occupational Safety and Health Administration (OSHA) requires employers to provide safe working environments, doing so isn't easy in construction, mining, law enforcement, firefighting, and the military. Each environment has risks and hazards that cannot always be managed or

CASE STUDY

Desert, Jungle, and Winter Mountain Warfare

The Marine Corps Physical Fitness Test (PFT) consists of a 3-mile (4.8 km) run, pull-ups or flexed hang, and abdominal crunches. Do these three items predict performance in the environments for which marines are trained?

In 1986 researchers were embedded with marines in several environments (the desert, the jungle, and the mountains in winter) during training to determine critical job tasks in each environment (Davis, Dotson, and Sharkey 1986). Working with experienced officers, they developed a test that included a sequence of the militarily relevant tasks. Marine performance was evaluated in each task, and scores were correlated to PFT scores and to laboratory fitness and body composition measures evaluated prior to the study. Correlation analysis was used to identify factors associated with performance. In the winter mountain environment, the analysis identified a leg endurance factor that related to performance with snowshoes, the run/walk, and sled pull tasks, and to the 3-mile (4.8 km) run. In the jungle environment, the researchers identified an arm endurance factor associated with paddling, swimming, and climbing/rappelling.

ⅢⅢⅢ LOW BACK INJURIES ⅢⅢⅢ

Ergonomics researcher Stover Snook identified three ways to reduce the incidence of industrial back injuries: (1) redesign demanding jobs, (2) use preemployment tests for demanding jobs, and (3) provide education and training (Snook 1991). Redesign is difficult in many demanding occupations and the military. Education and training make good sense, but the education is not that successful unless accompanied with muscular fitness training. Few employers require muscular training, but those that do report success in material handling tasks such as those encountered by soldiers (Williams, Rayson, and Jones 2002). As muscular strength and endurance improve, the incidence of low back and other injuries declines.

Reprinted from Jukkala and Sharkey, 1995 [unpublished research].

controlled (see the case study above). Workers often have to depend on their coworkers to provide assistance in emergencies. Job-related tests should ensure that the worker is capable of performing job duties, while also being able to meet expected or unforeseen emergencies that may require saving a victim or coworker.

Health

Low levels of fitness are associated with low back problems (Cady et al. 1979). Studies conducted on military recruits indicate that there are more injuries among low-fit recruits than among high-fit recruits (Knapik et al. 2003). Improved aerobic and muscular fitness reduces the risk for back problems and enhances the ability of men and women to perform militarily relevant tasks (Knapik 1997).

Low-fit workers are forced to work at a higher intensity, risking fatigue and injury. Workplace injuries cause loss of work time, reduce productivity, and add to the cost of workers' compensation. Failure to select and maintain capable workers leads to increased rates of job turnover and dramatic increases in employee injuries and can lead to health problems and degenerative disease.

Chronic fatigue also can contribute to health problems by suppressing immune function, which can lead to upper respiratory tract infections (Gaskill and Ruby 2004). Overwork and chronic fatigue are associated with increased levels of serum cortisol, which suppresses immune function. Those lacking the strength and stamina to perform on a daily basis are more susceptible to this condition than are those working well within their capacity. Fit workers have a substantially lower risk of overweight, heart disease, diabetes, and some cancers than low-fit employees have. We'll say much more about job-related physical fitness in chapter 12.

Many physically demanding occupations rely on annual or periodic medical examinations to maintain employee health. The National Fire Protection Association has medical requirements for structural firefighters (NFPA 1582), but the organization does not mandate annual work capacity tests or require participation in a job-related fitness program. Firefighters die of heart disease at a rate similar to that of the sedentary population. In 2002 the U.S. Fire Academy Firefighter Fatality Retrospective Study reported that heart-related fatalities accounted for 47 percent of deaths to structural firefighters, but only 7 percent for wildland firefighters (www.fema.gov). The report sited the exceptionally high fitness standards for wildland firefighters as one factor in the dramatic difference in heart-related deaths. Figure 4.1 illustrates the relationship of physical activity to the risk of heart disease. We will say more about medical issues in chapter 10.

Productivity

A well-designed work capacity test indicates an employee's ability to perform essential functions of the job. Wildland firefighters with higher test scores perform more fire line work throughout the day, but especially late in the shift, when compared with those with average scores. Performance is measured with electronic activity monitors (Gaskill and Ruby 2004). Career-long performance cannot be assured with a test that is administered only to recruits. Workers should be tested annually to ensure their ability to perform. Also, a validated work capacity test is an objective way to assess fitness for duty following an injury or prolonged illness.

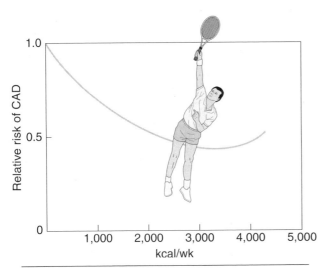

Figure 4.1   Physical activity and the risk of heart disease.

Reprinted, by permission, from B.J. Sharkey and S.E. Gaskill, 2007, *Fitness and health*, 6th ed. (Champaign, IL: Human Kinetics), 18.

Testing and Legal Issues

There are two categories of legal concern associated with job-related training and testing. One involves potential negligence or liability on the part of the department or agency associated with delivery of the training or the test. The other is potential negligence for not requiring training and a test. Indecision about testing does not appear to be an acceptable response.

In addition to a challenge of illegal discrimination, a candidate could charge the organization with negligence as a result of an injury experienced during training or testing. Even worse, the organization could be held accountable for a death associated with training or testing. In a strenuous occupation, such as firefighting, the risk of sudden death is associated with a lack of physical activity, gender (male), age (over 45), cigarette smoking, and overweight. Many firefighters who die of a heart attack show evidence of symptoms (elevated cholesterol, high blood pressure) and prior cardiac history. Surprisingly, only about 10 percent of all heart attacks occur during exertion.

When inactive people engage in vigorous exercise, their cardiac risk rises 56 times above that at resting levels, whereas the risk for the active person rises to a level slightly above the risk of sedentary living. The risk for active workers at rest is 50 to 70 percent lower than the risk of the sedentary population (Siscovick, LaPorte, and Newman 1985). For this reason we recommend that inactive candidates become active before they engage in training or take a test. For sedentary people that could involve six to eight weeks of walking, just to get ready to train. In chapters 10 and 12 we discuss medical standards for candidates and incumbents and outline standards for the delivery of training and testing programs.

An employer also could be susceptible to litigation for not requiring training and testing. The employer could be sued for hiring candidates who are not physically capable of performing safely, or for failing to require training of candidates or incumbents. For example, the Washington, D.C., Police Department was found negligent for not requiring a fitness program for an officer (*Parker v. District of Columbia*, 850 F.2d 708, 712 (D.C. Cir. 1)). The officer had been on light duty for four years because of an arm injury. During that

The BC Forest Service

In chapter 2 we discussed a court challenge of a test that went all the way to the Canadian Supreme Court. Although the court did not reject the need for a test, it ruled that the British Columbia government had not shown that it was reasonably necessary to the accomplishment of the government's general purpose, which is to identify forest firefighters who are able to work safely and efficiently. The BC Forest Service reviewed its options: develop and validate a new test, use an existing test, or stop testing altogether. It decided to adopt the job-related test used in the United States, the pack test. The original BC test required an aerobic score of 50 ml · kg^{-1} · min^{-1}, whereas the pack test predicts 45 ml · kg^{-1} · min^{-1}. In the first year of use in BC, 98.6 percent of the male recruits passed the pack test, as did 100 percent of the female recruits. Clearly, the test did not have an adverse impact (Bachop 2001).

The BC Forest Service also uses a validated muscular fitness test, the pump-hose test. The test requires a candidate to carry a 65-pound (29.5 kg) pump nonstop for 100 meters (no time limit). A timed portion of the test requires the candidate to carry a 68-pound (30.9 kg) rolled hose 300 meters and then drag a water-filled hose 200 meters (50 meters out and back, twice). This portion of the test must be completed within 4 minutes and 10 seconds.

time he did not participate in training. When he returned to full duty, he was assigned to arrest a person under warrant for an armed robbery. When the subject resisted arrest, the officer was unable to subdue him, lost control, drew his weapon, and used deadly force (www.ipmba.org/newsletter-0210-phys.htm).

An employer could be found negligent for failure to provide training, inadequate supervision during training, or not reassigning an incumbent who is not capable of meeting the demands of the job. A lawyer friend has said that you can be sued for many things; the goal of this book is to make sure the suit is not successful.

Test Development

This section covers two important steps in test development, the job task analysis and the development of job-related tests. Chapter 5 deals with the process of test validation, as well as reliability and objectivity. Chapter 6 addresses test implementation.

Job Task Analysis

A job task analysis (JTA) is one way to determine the work tasks the employee is required to perform on the job and to establish standards of performance. Another way is a carefully compiled list of essential functions established for the occupation by a panel of nationally recognized authorities. Both approaches should yield similar results. The advantage of the essential function approach is that the list covers all members of the occupation, not just one department. Because firefighters across the country use similar procedures and equipment, it should be possible to develop a national set of essential functions for the occupation. However, this is not the way it is currently done; at the present time, each department is expected to carry out its own process of test development. The same is also true of law enforcement. We describe the development of the JTA in this chapter and cover essential functions in chapter 16.

For hard work the JTA is used to determine important or critical tasks that define the physical demands of the occupation. There are a number of ways to conduct a JTA, including preparing a job description, observing employees performing job-related tasks, interviewing workers and supervisors (subject matter experts), designing a rating form, and using real-time measurements. In this section we describe several approaches that may be used to identify these important tasks.

We recommend that the JTA differentiate among the intensity, duration, and frequency of the task. It is also essential to identify tasks that are important or critical for safe and effective performance. Job-related tests should not be based on strenuous but seldom performed tasks unless the tasks are important or critical. Tests should bracket those unique dimensions that define the job and distinguish its necessary tasks from others.

Job Description

The job description usually describes what is expected of employees in general terms, often lacking the specificity that would lend itself to a job task analysis. But it is an excellent place to begin the job task analysis.

Observation

The job analyst can get a reasonable perspective on the nature of jobs performed in a limited number of locations by observing workers. For example, workers in the plywood industry are required to sort and orient sheets of veneer as they come down a conveyor belt. This repetitive task has a physical performance requirement that can be measured according to loads, frequency of lifts, and the

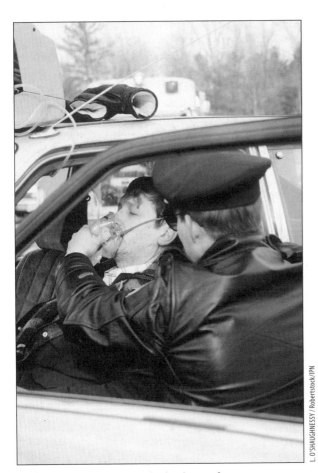

Observation may not be ideal in law enforcement, where critical tasks occur infrequently.

L. O'SHAUGHNESSY / Robertstock/IPN

energy cost of the task. The physical dimensions of the job can be replicated, but the ability to sort and other necessary physical skills require training and practice. Observation may not work as well in firefighting, which involves a number of tasks performed in a variety of environments, or in law enforcement, where critical tasks occur infrequently. This is when the interview becomes useful.

Interviews

The interviewing process is an effective way to determine the work tasks employees are required to perform and to establish standards. The process may begin with an informal interview of experienced employees and expand to a larger sample of employees. One approach is to use the Delphi technique. This multistep approach uses worker input to generate a list of potential tasks, refines the list in a group session with subject matter experts (knowledgeable incumbents and supervisors), and then solicits the input of a broad range of incumbents.

For example, in the St. Paul fire department case (Human Rights Commission 1989), Davis and Gilman conducted a comprehensive job task analysis for the position of firefighter using a department-wide survey. The 184 incumbent respondents identified critical, arduous, and frequently performed job tasks that provided the basis for the development of the criterion task test (CTT), a five-item test based on the JTA. The pace for a passing performance on the CTT was derived by having 10 supervisory personnel (district and deputy chief officers) independently view videotapes displaying a range of nine different paces on three different videotapes and indicating acceptable levels of performance.

A survey instrument based on management and worker input may be criticized as biased if sampling is not adequate. Sample size will depend on the size of the organization. A sample of 100 would be adequate for a small city police department, but would be too small for the New York City police department or the U.S. Marine Corps. Sample size could depend on the number of discrete tasks performed on the job. If workers do just one task, a large sample wouldn't be necessary.

JTA for Wildland Firefighters and Structural Firefighters

Wildland Fire. In 1973 five U.S. federal land management agencies and some states conducted a JTA for wildland firefighters. A wide range of firefighters were asked to list important and physically demanding firefighting tasks. The list was refined by subject matter experts from each agency, and the refined list was submitted to hundreds of experienced firefighters. The firefighters rated each task for importance (criticality), intensity, duration, and frequency. The top-rated tasks were as follows:

- Using hand tools to construct a fire line
- Hiking with light loads
- Lifting and carrying loads (such as a fire hose)
- Packing heavy loads

In 1975 the federal agencies adopted a work capacity test for wildland firefighters, a five-minute step test to estimate aerobic fitness. The test standard ($\dot{V}O_2$max of 45 ml $\cdot$ kg^{-1} $\cdot$ min^{-1}) was determined by doubling the average energy expenditure of firefighting tasks. That test was used until 1998, when a job-related test was adopted (Sharkey 1977).

To begin the transition to job-related tests for wildland firefighters, the JTA was revised in 1995 to reflect possible changes in the demands of the job. A survey or rating form with the original tasks and space for additional tasks was sent to a range of firefighters and fire managers. Respondents were asked to rate the original list and to add and rate additional entries. The highest-rated tasks were placed on a rating form and sent to several hundred fire personnel throughout the country (see Sample Rating Form for Job Task Analysis below). Ideally, surveys should be piloted (sent to a smaller select group) to ensure the relevance and clarity of the questions and format before they are distributed to respondents.

The tasks and categories on the rating form were explained in detail. For example, performing under adverse conditions, identified in interviews with experienced fire managers, was defined as "including long work shifts, rough steep terrain, heat, cold, altitude, smoke, insufficient food, fluid replacement, sleep." Emergency responses were defined as "fast pull-out to safety zone, rescue or evacuation assistance to others." Respondents were asked to rate each task, with 5 being high and 1 low. To emphasize important tasks, the importance rating (e.g., 4) was multiplied by the sum of the other ratings (add 5, 3, and 1 and multiply by 4 [$9 \times 4 = 36$]). The job task analysis confirmed the importance of aerobic fitness and the need for muscular fitness, and guided the test development process.

Sample Rating Form for Job Task Analysis

Rate the importance and physical demands of the following tasks: 5 is high, 1 is low.

Task	Importance	Intensity	Duration	Frequency
Using hand tools to construct fire line	_____	_____	_____	_____
Hiking with light loads	_____	_____	_____	_____
Lifting and carrying loads	_____	_____	_____	_____
Packing heavy loads	_____	_____	_____	_____
ADDED TASKS				
Performing under adverse conditions	_____	_____	_____	_____
Emergency responses	_____	_____	_____	_____
Chainsawing	_____	_____	_____	_____

Job Task Analysis for Wildland Firefighting. MTDC Project Record, Missoula Technology and Development Center.

Structural Fire. In another example, researchers developed a job-related test for structural firefighters. The test incorporates actual elements of the job identified in the JTA, including carrying hoses up several flights of stairs, raising hose rolls with a rope, chopping, pulling a charged hose, and carrying a victim. Based on the JTA, the test clearly represents the physical demands of the job (see table 4.1). Many departments have validated this or a similar test to select candidates for the job of structural firefighter (Dotson, Santa Maria, and Davis 1976).

Table 4.1 shows the relationship of fitness variables to performance, as indicated by test time. Note that excellent firefighters perform almost twice as fast as poor ones. Unfortunately, few departments require their firefighters to pass the test annually. However, the test has evolved into the popular Firefighter Combat Challenge, an international competition for firefighters.

Table 4.1 Firefighter Fitness and Performance

	EXCELLENT	GOOD	AVERAGE	FAIR	POOR
Age	**28**	**28.3**	**31.4**	**37.9**	**46.1**
Height (in.)	68.9	69.3	69.3	70	69.8
Weight (lb)	171.5	177.9	181	191.6	206.1
LBW*	145.5	145.1	143.1	141.5	146.2
% fat	14.8	18.1	20.3	25	28.5
Fat weight	26	32.8	37.7	50	59.7
Grip strength	116.6	107.8	105.1	101.3	95.1
Chin-ups	9	7.3	5.2	4.2	0.9
Push-ups	28.3	23.3	17.7	15	10.2
Sit-ups	47.9	45.9	35.1	29.5	22.8
$\dot{V}O_2$max	45.2	43.1	39.2	35.2	34.9
Test time**	5.68	5.73	6.84	7.88	10.75

*LBW = lean body weight

**Test time in minutes and decimal fraction of minute

Real-Time Measurements

If appropriate, an excellent way to analyze the elements of a job is through an ergonomic or metabolic analysis. If the task is repetitive and performed by most new-hires, it can be defined with lifting and carrying measurements or a metabolic analysis of incumbents, along with heart rate and lactate measurements. This information is useful in the process of moving from the job task analysis to test development.

There are many ways to conduct a job task analysis. To pass legal scrutiny and to meet the guidelines for employee selection procedures, the JTA should involve members of the department or agency in the process. Ultimately, the goal of the JTA is to provide an indisputable linkage to the test that will serve as the selection instrument. It focuses attention on demanding tasks, those critical, arduous tasks that define the mission. If employees must be able to lift or drag a victim out of harm's way, the employer does not need to determine whether they can carry a briefcase. Once the JTA has been completed, it is time to explore methods of test development.

An excellent way to analyze the elements of a job is with an ergonomic or metabolic analysis.

Photo courtesy of Brent Ruby

IIIIII **SMOKE JUMPER TEST** IIIIII

In the 1960s, well before the Uniform Guidelines on Employee Selection Procedures (2006) were formulated, smoke jumper managers developed a physical fitness test battery for this elite group of wildland firefighters. Required annually for recruits and incumbents, the battery includes a 1.5-mile (2.4 km) run to predict aerobic fitness (11 minutes = $\dot{V}O_2$max of 48.5 ml $\cdot$ kg^{-1} $\cdot$ min^{-1}), 7 pull-ups, 25 push-ups, and 45 sit-ups. The pull-ups were employed because jumpers have to be able to pull up and support their weight to release lines when they land in a tree! This long-standing test, which was approved by the Civil Service Commission, is supplemented by job task training requirements, including the 3-mile (4.8 km), 110-pound (50 kg) pack-out, which indicates the ability to pack one's gear out of the wilderness.

Type of Test

Some researchers prefer physical fitness tests, such as aerobic fitness or muscular fitness predictors of workplace performance. They believe that job-related tests do not discriminate or predict well and do not measure fitness (www.cooperinst.org). Others believe that fitness tests have questionable predictive validity with respect to specific occupational tasks (Jackson 1994). We do not engage in the debate because the laws and the courts have found that job-related tests have content validity and are therefore more credible than fitness tests. We use laboratory or field fitness measures to better understand the demands of a task, but do not recommend using fitness constructs in a job selection test.

Several questions arise in the process of test development: Should one task or several be used? That may depend on the number of important tasks and the degree to which they are correlated. If two tasks are highly correlated, one test may be sufficient. Multiple regression analysis will identify the tests that best predict success. Should candidates rest between tasks or go directly from one to the next? That could depend on how the

tasks are done in real life. If they occur infrequently, the test should allow time for recovery, but if the arduous tasks are done throughout a long shift, the test should mimic job demands. A physiological analysis (oxygen intake, lactate, muscular demands) of the proposed test will answer some questions. If the aerobic and muscular demands greatly exceed the effort used on the job, the test may be too arduous, it may cause adverse impact, and it is much more likely to be challenged.

Test development is an art as well as a science. Well-designed tests are viewed as challenges that successfully discriminate between those who can

Criteria in Test Development

- Use subject matter experts.
- Use incumbents in evaluation trials and seek their feedback.
- Include people of varying races and ages and of both genders in lab and field trials.
- Select tasks that are correlated to job performance.
- Conduct an ergonomic and physiological analysis of the test.
- Avoid high-skill tests that require training for recruits.
- Consider the costs of tests in time and money.
- Consider the cost of test equipment and the time to calibrate the equipment.
- Maintain records of the process.

and those who cannot do the job, without regard to race, color, sex, or age. The criteria in test development list on page 45 summarizes the key factors involved in test development.

In chapter 5 we show why we—and the courts—believe that job-related tests have content validity and are more credible than physical fitness tests as predictors of performance. Juries and judges look at a 1.5-mile (2.4 km) run or a push-up and see little relationship to the workplace. We know that the 1.5-mile (2.4 km) run is an excellent predictor of aerobic fitness for runners, and that the push-up is a test of muscular endurance for specific muscles. We also know that exercise and training are specific, and that this specificity cannot be ignored in training or testing. If running is an important work task, as it may be in law enforcement, then an employer should consider the running test (see figure 4.2). If not, then a job-related test is more appropriate.

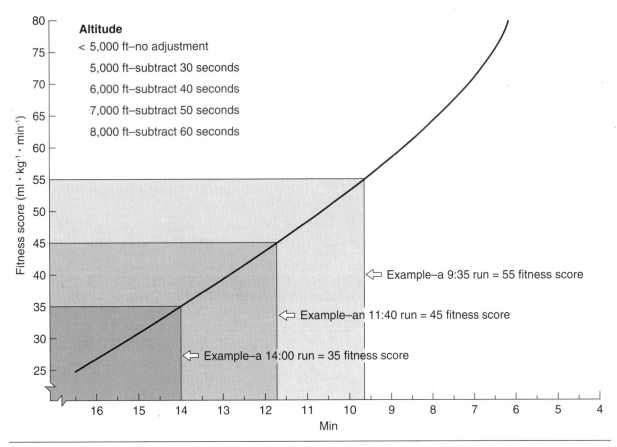

Figure 4.2 1.5-mile (2.4 km) aerobic fitness test. Note: Subtract altitude adjustment from 1.5-mile (2.4 km) run time. Then use the graph to find your score.

Reprinted, by permission, from B.J. Sharkey, 2002, *Fitness and health*, 5th ed. (Champaign, IL: Human Kinetics), 84.

Test Selection

In 1975, before the adoption of the current version of the U.S. federal Uniform Guidelines on Employee Selection Procedures (1978), and with the approval of the Civil Service Commission (now the Office of Personnel Management), federal land management agencies adopted an aerobic fitness test for wildland firefighters. Although the submaximal step test scores were correlated to the laboratory measure of aerobic fitness, subsequent field studies found that the correlations to measured work performances ($r = .5$ to $.6$) accounted for only 25 to 36 percent of performance on the tasks (r^2). A test that included aerobic and muscular characteristics was needed.

A large field study investigated candidate tasks for a job-related test. A total of 120 firefighters, representing the racial, ethnic, gender, and age characteristics of the firefighter population, volunteered to participate. After a battery of fitness and body composition tests, the subjects participated in six timed job tasks, including constructing a fire line with a hand tool, shoveling dirt, throwing dirt, deploying a fire hose, pulling a charged hose, and packing a load.

All the tasks were highly correlated with each other; performance on one task was significantly related to performance on the others (see table 4.2). The tasks with the best correlation to upper-body and lower-body performance and to total work (sum of all tasks), construction of a fire line with a hand tool and packing a heavy load, were selected for further study. The two tests were evaluated for their relationship to aerobic and muscular fitness, for validity and reliability, and for adverse impact. Although the female pass rate on the upper-body

line-building task was less than 80 percent of the male rate, the pack test did not have adverse impact, according to the Equal Employment Opportunity Commission (80 percent) definition. The line-building task was dropped because of adverse impact and for the time and cost of annual administration, nationwide, to thousands of firefighters.

A job-related test for wildland firefighters includes tasks such as working with a Pulaski axe, which is used both as a chopping axe and as a grubbing tool to dig fire line.

© John Storey/Time & Life Pictures/Getty Images

Table 4.2   Correlation Matrix

Task	Pack	Throw	Shovel	Hose lay	Charged hose	Line dig
Pack	1.0	0.65	0.64	0.79	0.68	0.69
Throw		1.0	0.80	0.56	0.86	0.77
Shovel			1.0	0.60	0.69	0.72
Hose lay				1.0	0.63	0.69
Charged hose					1.0	0.73
Line dig						1.0

$N = 120$ wildland firefighters. All correlations are significant: $p < 0.01$.

Laboratory tests indicated that the energy expenditure on the pack test was the same as the energy cost of the job, 7.5 kilocalories per minute (or 22.5 ml · kg^{-1} · min^{-1}) when the 45-pound (20.4 kg) pack was carried at 4.0 miles per hour. The test was standardized at 3 miles in 45 minutes to demonstrate the ability to sustain the energy expenditure required by the job. Completing the pack test in 45 minutes predicts a $\dot{V}O_2$max of 45 ml · kg^{-1} · min^{-1} (Strickland and Petersen 1999), the aerobic standard in place since 1975. Pack test performance was significantly correlated to upper- and lower-body measures of muscular strength (DeLorenzo-Green and Sharkey

1995; Sharkey, Rothwell, and DeLorenzo-Green 1994). The pack test was administered to 320 volunteer firefighters from several federal agencies and states, representing the racial, ethnic, gender, and age characteristics of the firefighter population. There was no adverse impact on the protected groups. The test was then administered to over 5,000 firefighters, again without evidence of adverse impact. The fire directors of five federal agencies adopted the pack test for servicewide use in 1998. The test is used by five federal land management agencies, a number of states, and fire organizations in British Columbia and Australia.

Summary

This chapter discussed the reasons for developing job-related tests and identified legal concerns such as liability and negligence. Employers should not let legal concerns prevent them from achieving the safety, health, and productivity benefits that come with well-designed and administered tests. Additional health benefits come when employers decide to test employees annually and include a job-related physical maintenance (fitness) program.

Real-world examples of job task analyses and test development were included to help you understand the test development process. Best results occur when incumbents and experienced managers are involved in the test development process. Their involvement avoids big surprises and grievances when the time comes to test all personnel. If possible, employers should conduct a physiological analysis of the job tasks as well as the test. For more on test development, see appendix B. For a chief's view of the process, see appendix C.

Chapter 5

Test Validation

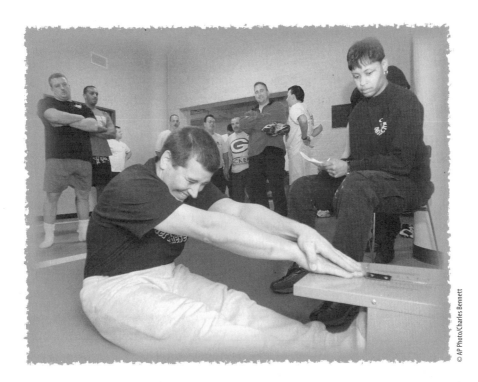

© AP Photo/Charles Bennett

The use of a selection procedure that has an adverse impact on the hiring, promotion, or other opportunities of members of any racial or ethnic group or gender is considered discriminatory and inconsistent with the U.S. Uniform Guidelines on Employee Selection Procedures, unless the procedure has been *validated* in accordance with the guidelines.

The term *validate* has several meanings, and all are relevant to this chapter. They include "to make legally valid," "to confirm the validity of," and "to verify." In research, a valid test is one that measures what it purports to measure. The validity of an employment test refers to the accuracy with which the test identifies the ability to perform the work, as defined by the job task analysis.

The Uniform Guidelines mention three approaches to validation: content, criterion-related, and construct validity. In this chapter we describe each approach and provide examples. Then we provide arguments in favor of one of the methods.

This chapter will help you do the following:

- Differentiate among content, criterion-related, and construct validity.
- Select an appropriate validation strategy.
- Determine test reliability and recognize the need for cross-validation.
- Identify bona fide occupational qualifications.
- Establish job-related test standards.
- Understand the consequences of adopting suboptimal selection procedures.

Content Validity

Content validity is established by sampling the content of important work tasks, work requirements, or outcomes identified in the job task analysis (JTA). The success of this approach depends on the qualifications of the subject matter experts

CASE STUDY

Firefighter Tests

Canadian researchers have developed several versions of a structural firefighter fitness assessment that is administered at university laboratories. Events include a treadmill test of aerobic fitness, a simulated hose drag, a 200-pound (91 kg) victim drag, a 56-pound (25.5 kg) ladder lift, a five-flight stair climb carrying an 85-pound (38.6 kg) hose, a 50-pound (22.7 kg) rope pull, and a 40-foot (12 m) ladder climb. All events but the treadmill and ladder climb are performed wearing a 40-pound (18.2 kg) vest and 4-pound (1.8 kg) ankle weights to simulate turnout gear and the weight of the self-contained breathing apparatus (SCBA).

Another version involves a stair climb carrying 25 pounds (11.4 kg), a hose drag, an equipment carry, a ladder raise and extension, a forcible entry, a 165-pound (75 kg) victim rescue, and a ceiling breach, all while wearing a 50-pound (22.7 kg) vest to simulate turnout gear and SCBA. Both versions include a crawling search event. The tests must be completed within a maximum total time (International Association of Fire Fighters 2007).

The International Association of Fire Chiefs (IAFC) and the International Association of Fire Fighters (IAFF) developed the IAFC/IAFF Candidate Physical Ability Test (CPAT), consisting of eight task-based simulations. Ten major North American cities collaborated with the U.S. Department of Labor in the development of the test and a labor–management wellness-fitness initiative. Test elements include a hose drag, a victim rescue, a maze crawl, a forcible entry while wearing a 50-pound (22.7 kg) vest, and a stair climb with the vest and two 12.5-pound (5.7 kg) shoulder weights. Although the goal of the pass/fail test is to improve the quality of life of all uniformed personnel, only candidates take the test. The wellness–fitness program is nonpunitive; that is, there are no penalties for employees who fail to meet objective fitness standards (IAFF 2007).

and employees who developed the JTA. Test items based on job tasks are said to have face validity, an obvious relationship that is not lost on juries and judges. Evidence of content validity should consist of data showing that the content of the test is a representative sample of essential aspects of performance on the job. Content-based validity is appropriate for physically demanding occupations in which workers regularly perform measurable tasks. The multiple-item tests of important structural firefighting tasks are an appropriate example of content validity.

Criterion-Related Validity

Criterion-related validity is demonstrated when the selection procedure is predictive of, or significantly correlated with, important elements of job performance. If the job requires prolonged hard work, aerobic fitness ($\dot{V}O_2max$) may predict performance. For example, the first job-related test used by wildland firefighters was an estimate of maximal oxygen intake, as predicted by a five-minute step test or the 1.5-mile (2.4 km) run. If the job requires lifting and carrying heavy loads, tests of muscular strength and endurance may be highly correlated to performance. Jobs that require aerobic endurance and strength may be related to aerobic and muscular fitness.

The U.S. military and some occupations have used body composition measures (body fat, lean body weight, body mass index or BMI) as job-related criteria. However, the courts prefer measures of performance, especially when body composition measures are likely to have adverse impact. Although lean body weight (fat-free mass) is significantly related to performance in lifting and carrying tasks, the measure has a disparate impact on women. In addition, the military has found that BMI, the ratio of height to weight, has little relationship to injury rates and attrition. For this reason the military does not use BMI for recruit screening or as a proxy measure for aerobic or muscular fitness. If fitness is important, employers should measure it (Sackett and Mavor 2006).

It seems relatively straightforward to prove a test's validity by correlating test scores to actual work performance. That can be done in material handling, but what can be done in the military, firefighting, or law enforcement, where the actual work is difficult to measure and dangerous? In the case of wildland firefighting, researchers can't go

Sgt. Julian Martinez, military policeman, provost marshal's office, Headquarters & Headquarters Squadron, Marine Corps Air Station Miramar, participates in the integrated strength training drills. The new training is part of the functional fitness concept adopted by the Marine Corps to condition marines for combat.

||||||| LAW ENFORCEMENT ||||||| TEST

The Cooper Institute developed and validated a fitness test battery that accurately measures fitness areas important in law enforcement. The battery includes a 1.5-mile (2.4 km) run for aerobic fitness, a 300-meter run, a vertical jump, a 1RM bench press or maximum push-ups, and a one-minute sit-up test (www.cooperinst.org, 2004).

to the fire to measure performance, such as feet of fire line dug per hour, because it can be dangerous. Also, firefighters work in crews, so individual performance is hard to differentiate. One answer is to simulate actual job tasks, measure individual performance, and correlate the performance with the job-related test. However, it takes motivated workers to sustain a high level of performance in a simulated, contrived, or make-work situation. Some will work hard, and others may only go through the motions. Researchers sometimes motivate workers with competition or monetary rewards or use heart rate monitors or lactate measurements to indicate effort.

Other approaches establishing the relationship of a test to performance include subjective job performance ratings; worker output (units per hour); or related variables such as absenteeism, lost-time injuries, workers' compensation costs, or attrition. Ratings and performance can be correlated to test scores. Absenteeism and other variables can be associated with levels of performance on the test. In the U.S. military, which has data on many thousands of recruits, low fitness is associated with injury and attrition.

Construct Validity

Construct validity is demonstrated when the job-related test is correlated to constructs associated with successful performance. Fleishman developed a list of nine physical abilities related to job performance (Fleishman and Mumford 1988). They included four of strength, two of flexibility, and coordination, balance, and stamina. Hogan (1991) reduced the list to three components: muscular strength, aerobic endurance, and movement quality (a combination of balance, flexibility, speed, power, and agility). These constructs sound a lot like the aerobic (cardiorespiratory) and muscular fitness criteria mentioned earlier. The construct approach does not clarify the validation process for physically demanding occupations, and it does not convey the direct relationship of a test to job performance, especially to a judge or a jury.

Validation Options

The Uniform Guidelines do not require that an employer conduct its own validity evaluation. The employer can adopt the standards of another employer that has completed an analysis. The success of this approach depends on the degree of similarity between the two employers. Adopting another's standards may save time and money, but it is less defensible if a legal challenge identifies significant differences between the two employers. If challenged, the employer's defense is to provide evidence that the demands of the job are essentially the same as those of the employer whose standards were adopted. Another option is for employers to cooperate in the development of a test.

CASE STUDY

The Pack Test

The pack test developed for wildland firefighters is highly correlated to the performance on other firefighting tasks (see table 4.2). When the test was correlated to lower-body task performances, the correlation coefficient was .95; when correlated to upper-body tasks, the correlation was .72; and when correlated to total performance, the sum of all tasks, the correlation was .82, which accounted for 67 percent of the variation in performance (Sharkey 1981). In a study conducted in an independent lab, the pack test was significantly related to aerobic fitness ($r = .69$); a score of 45 minutes on the 3-mile (4.8 km) test is equivalent to a $\dot{V}O_2$max of 45 ml $\cdot$ kg^{-1} $\cdot$ min^{-1} (Strickland and Petersen 1999). Pack test scores were highly related to measures of muscular strength (.61 to .72) and a fire line construction test (.79). And the pack test is reliable, with a test–retest correlation coefficient of .92 (DeLorenzo-Green and Sharkey 1995). The pack test conducted on a level course was correlated to a hill version ($r = .87$), but the hill version had an adverse impact on females. A field evaluation of the flat version on 320 firefighters did not reveal adverse impact (Sharkey, Rothwell, and Jukkala 1996), and a larger trial of 5,000 firefighters confirmed that there was no adverse impact according to race, sex, age, or national origin. Five federal agencies, cooperating under the National Wildfire

A level-course pack test.

Coordinating Group, have adopted the pack test as the validated standard for wildland firefighters.

The ideal situation is to adopt a list of essential occupational functions established by a reputable organization, such as the National Fire Protection Association in the United States. The functions represent the job task analysis for the profession on a national level, eliminating the need for individual departments to conduct their own JTA and test validation.

Reliability and Cross-Validation

Test–retest consistency is an important factor in job-related testing. If a test is not reliable, if test–retest scores are dissimilar and correlations are low, the test cannot be viewed as a valid measure of performance. A rested person should get a similar score on the retest as on the test. To determine

test–retest reliability, the employer should select a representative sample and test on two different days. Correlate the two sets of scores and determine mean differences. If the correlation is high and the mean difference is small, the test is reliable.

A well-designed test is easy to score. Results should not depend on administrative decisions, distortions, or prejudices; the test should be objective. Effective instructions for test administrators and for candidates will improve test reliability and objectivity.

Training and practicing a test for a physically demanding occupation will improve candidate performance. Aerobic and muscular fitness training helped female U.S. Army recruits significantly improve material handling and load carriage, carrying 75 pounds (34 kg) at 4.4 miles (7 km) per hour (Harman et al. 1997). We'll say more about test administration in chapter 6.

To avoid overestimating the validity of a test, an employer should cross-validate it on a separate sample, that is, validate on one group and cross-validate on another. As before, the sample should mirror the characteristics of the workforce. Volunteer incumbents should be tested on the job-related test and the battery of work tasks. Employers should then correlate test scores with task performances and, if measured, with physiological criteria. Cross-validation involves administering the test to other incumbents, recruits, or incumbents from another department. In both cases the test should demonstrate content validity with the work samples and, if measured, criterion-related validity with related physiological variables.

Bona Fide Occupational Qualification

Valid and reliable bona fide occupational qualifications (BFOQs, or BFORs [bona fide occupational requirements] in Canada) provide a means of determining whether a person can meet the specific requirements of the job effectively and efficiently without undue risk of injury. One would assume that people interested in pursuing employment in physically demanding occupations, such as firefighting and law enforcement, would recognize the physical demands involved in such positions and physically prepare before applying to work. One also might assume that employees performing physically demanding work would appreciate the obvious need to maintain their work capacity. Surprisingly, this is not the case.

In an analysis of U.S. firefighter fatalities, Fahy (2005) noted that almost half of the firefighters who died on duty fell victim to sudden cardiac death. Steps to reduce the risk of heart attacks include conducting annual medical examinations, screening for coronary heart disease risk factors, conducting stress tests for those with multiple risks, treating risk factors, and restricting firefighters with positive stress tests. The analysis noted that 73 percent worked in departments that did not have a program to maintain firefighter fitness and health, as described in NFPA 1500, Standard on Fire Department Occupational Safety and Health Program. In rural communities, 88 percent of firefighters worked without these programs. Few of the departments that have fitness and health programs require participation or annual perfor-

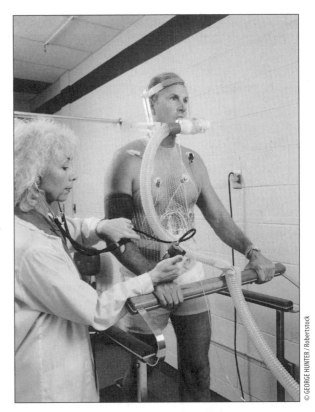

Steps to reduce the risk of heart attacks include annual medical examinations, screening for coronary heart disease risk factors, and stress tests.

mance testing. A more recent study found that firefighter deaths were associated with suppressing a fire (32.1 percent), responding to an alarm (13.4 percent), returning from an alarm (17.4 percent), physical training (12.5 percent), nonfire emergencies (9.4 percent), and performing emergency duties (15.4 percent). Based on the risk and the time involvement for each, fire suppression had a risk 10 to 100 times as high as that for nonemergency duties (Kales et al. 2007).

Absolute Standards

Should job or fitness standards be absolute—that is, the same for all employees—or relative to age and gender? Many organizations established separate standards for men and women and then regressed or age-adjusted the standards. Using relative or adjusted standards obscures the meaning of a standard or requirement. Hard work doesn't get easier for women, for smaller people, or with advancing age. If challenged in court, relative standards would be hard to defend. And yet, to avoid adverse impact, many organizations use relative standards or modified tests. We argue that a standard is a standard, and that those performing the same job should meet the same standard, regardless of gender or age, unless there is compelling evidence to the contrary.

Age

In chapter 2 we alluded to the effects of age on aerobic fitness, with a decline approaching 8 to 10 percent per decade for inactive people, regardless of their initial level of fitness. Those who decide to remain active can cut the decline in half (4 to 5 percent per decade), and those who engage in fitness training can cut that rate in half (2 to 3 percent per decade), at least for a while. Although trainability may decline somewhat with age, aerobic and muscular fitness can be improved, even after the age of 70.

Gender

Courts have contributed to the confusion concerning work standards by accepting the relevance of employment tests and then ruling against tests for adverse impact. A lumber company administered job-related strength and aerobic fitness tests, and the women's pass rate was 42 percent of the men's pass rate. Although the court noted that the company was able to demonstrate the job-relatedness of the tests, that the tests served the legitimate goal of hiring fit workers, and that those who pass are more likely to do the job adequately and safely than those who do not pass, the test "unnecessarily excludes qualified applicants, a disproportionate number of whom are women." The court ordered a decrease in the aerobic standard for women (from 35 ml · kg^{-1} · min^{-1} to 29.75) and a variable-height stepping bench (under 11 in. [28 cm]) for the aerobic test, based on the height of the female applicant (*EEOC v. Simpson Timber Co.* 1992).

The court created a separate test for women, ignoring their ability to train and meet the company's legitimate employment goal of a physically capable workforce. The average 40- to 49-year-old female, with a score of 32 ml · kg^{-1} · min^{-1}, could train and improve 20 percent to a score of 38 and easily meet the standard (see table 2.1, p. 15). Instead the company has two standards of employment, and women who pass below the original standard will be less capable and more prone to injury. With age, assuming that they do

---

|||||||||| ADVERSE IMPACT ||||||||||

A law enforcement agency established height and weight standards for new corrections officers (*Dothard v. Rawlinson* 1977). The standard was based on the relationship of size to strength and the ability to perform strenuous law enforcement tasks. The minimum height was 5 feet 2 inches (157 cm) and weight was 120 pounds (54.4 kg). The standard was found to have adverse impact because it excluded one-third of women but less than 2 percent of men. The U.S. Supreme Court ruled that if strength is a valid job requirement, then a direct test of strength should be used.

---

not engage in training, they will become even more marginal.

You may be wondering why we focus on age and gender and ignore race and national origin. Published research studies do not document work deficiencies in these protected classes. Moreover, the literature of work physiology does not identify differences due to race, ethnicity, or national origin. Studies on firefighters do not detect differences that lead to adverse impact within these groups (Sharkey 1981, 1997b).

Studies in work physiology do not suggest racial or ethnic differences in work capacity. With some doubting the validity of race as a useful classification and others downplaying the existence of meaningful racial differences, there is little justification for a lengthy discussion. What is clear is that job-related work capacity tests do not unfairly affect racial or ethnic groups. Indeed, valid work capacity tests can be considered color-blind.

Test Standards

How are test standards identified in the test development process? The answer depends on the relationship of test scores to performance, the type of validity employed, the degree and quality of supporting evidence, and other factors. Establishment of a minimal passing score (or cutoff score) depends on the available pool of labor, the need for productivity and worker safety, and the potential for adverse impact (Jackson 1994). The U.S. military has to balance a desire for performance with the pool of available candidates. If they set standards that are too high, they can't reach their recruitment goals. In the public and private sectors, setting extremely high standards is a recipe for adverse impact and legal challenge.

The federal Uniform Guidelines state:

> "Where cutoff scores are used, they should normally be set so as to be reasonable and consistent with normal expectations of acceptable proficiency within the work force. Where applicants are ranked on the basis of properly validated selection procedures and those applicants scoring below a higher cutoff score than appropriate in light of such expectations have little or no chance of being selected for employment, the higher cutoff score may be appropriate, but the degree of adverse impact should be considered" (Uniform Guidelines 1978).

This warning seems compelling in light of the decision regarding the *EEOC v. Simpson Timber C.* case discussed earlier.

Cutoff Score

The minimal acceptable score on a test should ensure performance now and in the future and minimize adverse impact. In the *Simpson* case, the court arbitrarily reduced the women's aerobic standard 15 percent to a level attainable by the average 50- to 59-year-old female, and undoubtedly harmed the mill's productivity. Henderson, Berry, and Matic (2007) showed that a low cutoff score reduces the utility of physical testing to near zero, and it will not reduce disparate impact on female applicants until the cutoff score is set near the male fifth percentile. Table 5.1 illustrates the options available with a test. The goal is to pass those who can do the work and fail those who cannot.

Unfortunately, no test is perfect: A few who can do the work will fail, and some who can't do the

CASE STUDY
Minimum Standards

A sample of 153 firefighters, selected for demographic characteristics, performed a number of firefighting tasks and a battery of physical abilities tests (Sothmann et al. 2004). Regression analysis revealed three predictor test items (hose drag or high-rise pack carry, arm lift, and arm endurance) that significantly predicted task performance time. The predictor tests correctly identified 89 percent of successful performers (specificity) and 72 percent of unsuccessful performers (sensitivity). Validation of minimum performance standards will identify people with a high probability of not being able to perform critical fire suppression tasks.

work will pass. Setting a low cutoff score minimizes failure by those who can perform, but it passes more who cannot do the work. Setting a high cutoff score increases failure by those who can do the work, but it eliminates those who cannot. A high cutoff score also increases the risk of adverse impact and legal challenge. An alternative is to set a cutoff score that is reasonable and consistent with normal expectations of acceptable proficiency within the workforce and then select the best performers who meet the standard.

The cutoff score should be set to ensure performance now and in the future. A barely qualified 20-year-old recruit is unlikely to excel in 10 or 20 years. It is not advisable to use the lowest score of currently employed workers to set the cutoff score, as some would suggest. Instead, subject matter experts should select the minimum acceptable score after viewing video recordings of varying rates of performance. The opinions of successful incumbents should also be factored in. The physiological cost of performance as well as research studies or other validated tests should also be used in the decision.

Determining an arbitrary cutoff with a statistical technique, such as one standard deviation below the mean scored by incumbents, is not advisable. That technique, which accepts 84 percent of the current employees, may be acceptable with a trained and fit workforce, but it may be a major mistake with an aging, inactive pool of incumbents. This approach, as advocated by the plaintiff's expert witness in *Lanning v. SEPTA* (2000), has little

Table 5.1   Possible Test Outcomes

	Can do work	Can't do work
Pass test	Pass and can do work* (true positive test)	Pass but can't do work (false negative test)
Fail test	Fail but can do work (false positive test)	Fail and can't do work** (true negative test)

*Test specificity: the percentage of correctly classified successful performers
**Test sensitivity: the percentage of correctly classified unsuccessful performers

or no utility in the selection of recruits. It results in a cutoff score unlikely to discriminate among candidates, one that passes most men. This approach does not ensure the ability to perform; instead it establishes a race to the bottom.

Ranking

Employers use a variety of measures to judge prospective employees, including written tests, personality assessments, medical evaluations, and even drug tests. They may factor performance in each category into a final decision. When the job involves physically demanding work, performance of job-related tests should be emphasized. Ranking candidates on the work test ensures this emphasis. Selecting top scores in rank or top-down order maximizes performance. Content-related procedures support rank ordering. Cutoff scores may be used to increase workplace diversity even though it may mean somewhat diminished performance. In dangerous occupations, such as law enforcement, firefighting, and the military, in which safety is critical, selecting top performers can be justified. The recommended approach is to rank those who exceed the cutoff score and select the most qualified applicants. The organization should document the rationale used for cutoff scores, ranking, or both.

Setting Scores

Maximal oxygen intake ($\dot{V}O_2$max) has been associated with performance in firefighting. Studies of wildland firefighters recommend a minimal aerobic fitness of 45 to 50 ml · kg^{-1} · min^{-1}. (Docherty, McFayden, and Sleivert 1992; Sharkey 1997b). Most studies of structural firefighting estimate the aerobic requirement between 40 and 45 ml · kg^{-1} · min^{-1} (Dotson, Santa Maria, and Davis 1976). Yet one group studied firefighters and found that the lowest possible aerobic fitness score was 33.5 ml · kg^{-1} · min^{-1}, a level attainable by the average male in the 60-to-69 age group and the average female in the 40-to-49 group (see table 2.1). This minimum score was based on the performance of 20 experienced firefighters performing firefighting tasks, but not involving rescue of a victim (Sothmann et al. 1990). The $\dot{V}O_2$max is a meaningful criterion for critical firefighting tasks, wildland or structural. Cutoff scores should reflect a level con-

sistent with performance, not the minimum score demonstrated by unfit, overweight employees. Recruits with a $\dot{V}O_2$max of 33.5 would struggle to perform demanding tasks and fail as age diminishes their already compromised capacity.

The wildland $\dot{V}O_2$max criterion was based on the measured energy demands of the work and the prolonged arduous work shifts. Because workers do not sustain more than one-half of their $\dot{V}O_2$max during an 8-hour shift, and wildland firefighters work 12- to 16-hour shifts, the energy cost (7.5 kcal or 22.5 ml · kg^{-1} · min^{-1}) was doubled to determine the minimum passing score (45 ml · kg^{-1} · min^{-1}) (Sharkey 1977). Structural firefighters may expend more energy per minute but for shorter periods, so the recommended minimal standard is 42.5 ml · kg^{-1} · min^{-1}. Higher minimum standards could be defended for performance and safety, but not without increasing the likelihood of adverse impact. Cutoff scores reflect a balance of factors, such as job demands, physiological capacities, safety issues, the labor pool, diversity needs, and more.

Muscular fitness scores may be used as criteria in the validation process. However, there is no standardized strength or muscular endurance test to guide the setting of cutoff scores. Years ago researchers used static or isometric strength measures, until they found that dynamic measures were better related to work performance. They used upper-body measures to represent muscular fitness, until they found out that performance was specific to muscle groups used on the job. Today some employ isokinetic measures of strength and muscular endurance. We recommend the use of job-related tasks (e.g., material handling) or published muscular fitness measurements that have been correlated to performance in the occupation to help define the minimum score. Extremely high cutoff scores are not advisable unless the employer is willing to offer job applicants pretest training.

A work-related approach can be used to determine the minimum acceptable test score. In the St. Paul Fire Department case (Human Rights Commission 1989), Paul Davis and Bill Gilman videotaped firefighters performing a series of work tasks. The cutoff pace for performance of the work test was derived by having 10 supervisory personnel (district and deputy chief officers) independently view videotapes displaying a range of nine paces and indicating acceptable levels of performance. A similar approach could be used for material handling and other tasks. In the St. Paul Fire Department case, Davis and Gilman also used oxygen intake and lactate measures to characterize the effort expended and the degree of fatigue. Using several approaches helps to bracket a reasonable score that is consistent with normal expectations of acceptable proficiency.

The validation checklist below summarizes the key factors involved in validating an employment test.

Validation Checklist

- Do a job task analysis.
- Identify and develop job-related tests.
- Conduct a physiological analysis of tests.
- Select a sample of incumbents that mirrors workforce demographics.
- Measure important physiological criteria of the sample.
- Test incumbents on the job-related test and a battery of work tasks.
- Correlate test scores with task performances and physiological criteria.
- Determine test reliability (see page 52) and adverse impact.
- Cross-validate the test on another sample.
- Determine content and criterion-related validity.
- Determine the cutoff score.

For further assistance in the development of job-related tests, consult the Uniform Guidelines on Employee Selection Procedures (1978), and the Principles for the Validation and Use of Personnel Selection Procedures (2003) developed by the Society for Industrial Organizational Psychology (www.siop.org).

Suboptimal Selection Procedures

Despite the substantial literature that argues persuasively for high selectivity in hiring, many agencies adopt suboptimal selection procedures, sometimes through lack of awareness and some-

times as a result of external pressures, including litigation. Most often the strategy includes reducing selectivity by setting a very low cutoff score on some troublesome component of the screening procedure, such as a physical abilities assessment, because of the large gender differences encountered. The effect of the low cutoff score decision is twofold. First, it shrinks the utility of the physical abilities component to near zero with respect to identifying the most physically qualified candidates, even when the validity of the screening procedure is high. Second, regardless of what might have been found in a job analysis or a criterion-related validation study, the relative contribution of physical scores to selection becomes trivial relative to the other screening components, calling into question the validity of the overall screening procedure (Henderson, Berry, and Matic 2007). For a detailed discussion of the effects of suboptimal selection procedures, see Dr. Henderson's analysis in appendix D.

Summary

This chapter defined validation and differentiated among content, criterion-related, and construct validity. Based on its acceptance in the courts, we recommend content validity for physically challenging occupations. Criterion-related validity is acceptable, but a combination of content and criterion-related validity may be the best solution. Content validity ensures that the test is job related, and criterion-related validity links the test to physiological criteria and capacities.

We mentioned the importance of test reliability in the hiring process, and the need for cross-validation was emphasized to minimize the risk of sampling errors. The chapter discussed bona fide occupational qualifications, absolute versus relative standards, and the influence of age and gender in physically demanding occupations. We suggested that neither age nor gender should disqualify those willing to meet and maintain objective standards. The chapter ended with thoughts on how to establish job-related test standards. Minimum, or cutoff, scores establish minimum standards: Choosing from the top down ensures the most qualified employees. Of course, job-related fitness will decline rapidly if it is not maintained in a year-round program punctuated with periodic (annual) tests of work capacity.

Chapter **6**
Test Implementation

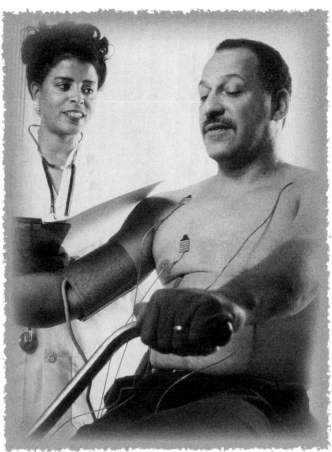

© Gabe Palmer/CORBIS

It is a challenge to develop and validate a job-related test for a physically demanding occupation, yet it can be even more challenging to successfully implement the test in the workplace. A number of factors can enhance the likelihood of success, but none is more important than early and ongoing employee involvement. The job-related test should be presented as a way to promote and support employee health and safety, to reduce injuries and illness, and to enhance physical and emotional health and morale. Establishment of the job-related test is an important step in the development of a comprehensive employee health program, consisting of medical standards and exams, entry-level testing, health education and health-related fitness, job-related fitness training, periodic (annual) evaluations of work capacity, and return-to-work testing after lost-time injury or illness (covered in part IV).

This chapter will help you do the following:

- Consider the need for professional assistance in test development and implementation.

- Understand important personnel issues involved in test development and implementation.

- Determine implementation strategies.

Seeking Professional Assistance

A question an employer needs to ask before developing and implementing a job-related test is this: Does the department or agency have the appropriate staff to develop, validate, and implement a job-related test? The employer should consider the need to conduct a job task analysis and appropriate statistical calculations to develop and validate the test and establish acceptable levels of performance. A thorough analysis of legal implications is also needed. Some organizations proceed with existing staff, some seek limited assistance, and others contract for professional services. Whatever the decision, the employer should assemble a working group composed of employees, middle managers, managers, a union representative, a medical advisor, and legal counsel. A professional may also be needed to guide or carry out the process.

Employers who doubt the need for professional assistance should read the transcript of the *United States v. City of Erie* (Pennsylvania) case, in which the court struck down a reasonable test because it lacked sufficient validation and proof of business necessity, items an expert would establish. The *Erie* case is discussed in more depth in chapter 7. For the time being it is enough for employers to understand that if they wait until a legal challenge to seek professional assistance, they and their experts will be fighting an uphill battle.

An Internet search for organizations that perform this work will reveal numerous options. Employers could contact similar departments that have developed tests or get advice from professional organizations or university experts. We advise employers to seek advice from an individual or organization with a record of performance in the field who understands the dimensions of physically demanding occupations. The appropriate area of study for hard work is exercise physiology, also known as applied physiology, work physiology, occupational physiology, or human performance. The physiologist should be an active member of the American College of Sports Medicine or a similar reputable organization with professional and research presentations and publications.

Employers should seek assistance from those who have successfully advised organizations similar to theirs. As they would with any contract, they should ask to see references, resumes, tests, and supporting materials. They could ask previous clients their opinions of the organization. And they should examine the consultants' success in defending their products in court. When hiring a consulting firm, employers should consider how its reputation and previous work may be viewed by a judge or jury. A knowledgeable consultant could help transfer a test or adopt the standards of another agency. However, the authority having jurisdiction will be responsible for the defense of the test and how it was developed.

The appropriate area of study for hard work is exercise physiology. A physiologist assisting an employer should be an active member of the American College of Sports Medicine or a similar reputable organization with professional and research presentations and publications.

Personnel Issues

Among the important personnel issues involved in test development and implementation are employee involvement, subject matter experts, management–labor negotiations, confidentiality, gender equity, and pretest training. Careful planning at this stage could help to minimize future problems.

Employee Involvement

It is important to involve employees early in the test development process. Employers should seek their input on the job task analysis, conduct early test trials on volunteers, and use their performances to set passing scores. More importantly, employers will want incumbents to have ownership in the process so they can see how they are doing and someday agree to implement the test. As they work toward the immediate goal, employers should let employees know departmental aspirations, assuring them that each step depends on employee approval.

Subject Matter Experts

Employers should take advantage of the knowledge of experienced middle managers in the test development process. They should survey incumbent experts to determine physically demanding field tasks, then use them and other employees to analyze and refine the JTA. The knowledge of these experts can be used to determine the minimum or cutoff score from published studies, video recordings, or some other approach. It is very important for all employees to have ownership in the process, especially those who will be administering the program. If experienced subject matter experts are not available, employers can seek assistance from another department similar to theirs.

Management

Managers must be convinced that the program will ensure the employment of those capable of performing efficiently and safely, and that dollars spent on the program will be returned in the form of higher productivity and a safer working environment. It is essential that the captain, chief, or top manager believe in the program and that he or she be an outspoken advocate, even an example. That doesn't mean that the manager has to take the test, but he or she should demonstrate personal involvement in a health and fitness program. Programs with sincere management involvement are far more likely to succeed.

Labor Negotiations

If a labor contract is in effect, it is essential to involve the union in test development and implementation from the beginning. Unions have been receptive to entry-level tests, but less so to periodic (annual) tests or mandatory job-related fitness programs. However, as professionalism and performance grow in importance, unions may be more receptive to measurable indexes of productivity. Step 1 is to implement the entry-level test. Future steps can be negotiated by linking performance to salary increments and other inducements, such as fully compensated medical insurance.

Confidentiality

A crucial factor in job-related testing of recruits or incumbents, especially since the passage of the Health Insurance Portability and Accountability Act (HIPPA) in the United States, is the need for confidentiality in all aspects of the program. HIPPA requires that employers safeguard the security and privacy of health data and establish national standards for electronic data transfer. If an applicant doesn't receive medical clearance before a test, the reason cannot be shared with others without a release from the applicant. Test results and personnel actions must be kept confidential.

Gender Equity and Pretest Training

To ensure gender equity in recruitment for physically demanding occupations, many organizations offer practice or opportunities for pretest training (see the case study, Preconditioning for Basic Combat Training). The application of nondiscriminatory test standards in the hiring and retention of public safety officers has become controversial when gender differences in average levels of aerobic and muscular fitness lead to adverse impact. One approach is to provide training to minimize the differences.

Preconditioning for Basic Combat Training

A large study (Knapik et al. 2006) evaluated the impact of increasing the fitness of low-fit U.S. Army recruits before they enter basic combat training (BCT). When recruits took the entry fitness test, those who failed were assigned to either special preconditioning (PC) fitness training or no preconditioning (NPC). PC recruits entered BCT when they passed the entry test, whereas the NPC recruits went directly into BCT. The study showed that low-fit male and female recruits who underwent preconditioning before BCT had reduced attrition and tended to have lower injury rates when compared with recruits of similar low fitness who did not precondition. Of course, recruits who passed the initial fitness test and went directly to BCT had even less attrition and risk of injury.

© Human Kinetics

To ensure gender equity in recruitment for physically demanding occupations, many organizations offer practice or pretest training.

Physical Work Capacity and Gender Differences in Law Enforcement

Shephard and Bonneau (2002) summarized the impact on physical work capacity of commonly encountered gender differences in body size, body composition (lean body weight, or LBW), endurance, and muscular strength for police officers. The authors summarized employment laws, criteria for establishing a bona fide occupational requirement, and the development of standards that satisfy the criteria. Requirements are based on the task to be accomplished. Contrary to the statement in the *Mejorin* decision (British Columbia, see page 17), "owing to physiological differences, most women have a lower aerobic capacity than most men and that, unlike most men, most women cannot increase their aerobic capacity enough with training to meet the aerobic standards," the potential training response of female applicants is likely to at least match that of their male peers, and the needs of female police recruits are best accommodated by providing every opportunity to augment fitness to the demands of the job. The authors pointed out that the main weakness of any current requirement is that most police departments do not yet apply an equivalent evaluation to older incumbent officers, for whom similar fitness issues exist.

Kramer and associates (2001) studied the effects of resistance training programs on strength, power, and military occupational task performances in women. Ninety-three untrained women were tested for body composition, strength, power, endurance, repetitive box lifts, a 2-mile (3.2 km) loaded run, and U.S. Army Physical Fitness Tests, before and after six months of training. Strength training improved their physical performances. Upper-body and total-body resistance training resulted in similar improvements in occupational tasks, especially those that involved upper-body musculature. Gender differences in physical performance measures were reduced with resistance training, which underscores the importance of this training for physically demanding occupations.

Until women are given the chance to work up to defensible job standards, we'll never know how good they can be. Employers should establish valid job-related tests, advertise the standards, and, if possible, provide preemployment training. If few women qualify, employers should not panic and lower the standards. Rather, they should hold firm and watch as women come forth to accept the challenge.

Implementation Strategies

Law enforcement, firefighting, and military organizations expect their applicants, trainees, and incumbents to maintain an adequate level of fitness for duty. This section outlines strategies for the implementation of the job-related test. We begin with a test for recruits, then move to strategies that include incumbents. Fitness should be maintained throughout a person's career if the job requires the ability to respond physically.

Recruits and Incumbents

Most law enforcement and firefighting organizations require applicants or trainees to pass a job-related test. Thereafter, incumbent employees are never again evaluated for work capacity. It is relatively easy to introduce a test for recruits, but difficult indeed to require one for incumbents. For this reason, the first step is to introduce the entry test.

Once the test for recruits is up and running, employers can offer to test interested incumbents so they can see how they score relative to recruits. They could also schedule a recruit versus incumbent event to foster friendly competition. When all parties agree, the employer can begin to phase in incumbents, providing ample time and assistance to prepare for the test. The test could be delivered annually or periodically (e.g., every two or three years). However, more frequent testing (e.g., semiannually) is likely to foster an ongoing interest in training.

One way to diminish the resistance to incumbent testing is to offer to phase the program in with age adjustments or to "grandfather" older employees. Although we don't advocate age adjustments for hard work, they do help to alleviate the concerns of older employees. "Grandfathering," or not requiring the test for those already in the job, should be reserved for those who are close to retirement or reassignment. As employees move from being recruits to being seasoned veterans, many move to less physically demanding administrative positions. There is no need to require a work test

for managers who no longer perform physically demanding duties.

When the test is administered to incumbents, what should an employer do with an employee who cannot or will not meet the standard and is not suitable for reassignment? The agreed-on written test requirements must cover this possibility. Some departments reassign the employee to less demanding work and provide time to train and meet the standards. After six months or a year, the employee may be let go if he or she fails to pass the validated job-related test of work capacity.

Managers

Although managers seldom do the hard physical work performed by other employees, they do need a certain level of fitness for performance and health. Any sworn law enforcement officer, regardless of rank, should intervene when a felony is committed in his or her presence. The department or agency should consider their needs as it continues to develop the program. Managers should not be required to take a vigorous test, however, unless they are active and have engaged in appropriate training.

Even managers need a certain level of fitness for performance and health.

Return to Work

The job-related test provides a way to measure a recruit's or incumbent's work capacity. For example, when U.S. federal land management agencies were developing a test for the arduous work of firefighting, they decided to develop job-related tests for arduous, moderate, and light work to cover a number of incident command and fire camp jobs. To test all three levels at the same time, the test was modeled after the pack test. The moderate category (see page 15) covers less arduous fieldwork, including safety officers who walk several miles with light packs while performing their duties. The light category is intended for those who work in the fire camp, which occasionally has to be moved—quickly—when the fire makes an unexpected run (see table 6.1).

A job-related test can also be used to assess a person's ability to return to work after a lost-time injury or illness. If an employee can't perform the test, perhaps he or she is not ready for the workplace. A well-designed test provides a way to train for the return to work. Train? you ask. Yes, train, just as athletes do after an injury. They begin with rehabilitation, and when full function has returned, they engage in sport-related training to prepare for a return to competition. If they go back too soon, they risk aggravating the injury and losing more time. Motivated athletes and employees want to return to work as quickly as possible. Employers should provide training and a test to be certain they are ready. Of course, employees should always seek medical approval before a return to duty after a lost-time injury or illness.

Summary

In this chapter we considered the need for professional assistance in the development and validation of a job-related test of work capacity. We also provided suggestions on how to seek that assistance, if needed. Although a number of professions engage in written and skill test development, they don't emphasize job-related performance standards or the need to train for physically demanding occupations. Members of the Society for Industrial and Organizational Psychology understand tests of knowledge, skill, and abilities, but they have less experience with the physiological requirements for performing hard work. The goal of ergonomists is to adapt the job to the worker. Exercise physiologists understand hard work, and they know how to adapt the worker to the job. We recommend that employers seek a professional with ties to the American College of Sports Medicine and its emphasis area in occupational medicine and physiology (for more information, see www.acsm. org/).

This chapter also addressed the need to include all levels of personnel in the test development process and emphasized the importance of management approval and personnel involvement. Early participation by the union is essential in a collective bargaining unit. Employers should negotiate an agreement with the union and, if possible, develop a phase-in program for incumbents.

Pretest training is an affirmative action that could help reduce the likelihood of adverse impact. Until women are given the chance to work up to defensible job standards, we'll never know how good they can be. Employers should establish valid job-related tests, advertise the standards, and provide preemployment training programs.

Test implementation strategies were discussed for recruits and incumbents. Although the test is important, it isn't the ultimate goal. The real key to safety, health, and productivity is an employee health and wellness program that includes medical surveillance, periodic tests of work capacity, and job- and health-related fitness, along with periodic medical tests and health and wellness education. We describe these components in chapters 10 and 12.

Table 6.1 Wildland Firefighting Work Capacity Tests

Category	Test	Length	Load	Time	$\dot{V}O_2$max*
Arduous	Pack test	3 mi (4.8 km)	45 lb (20 kg)	45 min	45
Moderate	Field test	2 mi (3.2 km)	25 lb (11 kg)	30 min	40
Light	Walk test	1 mi (1.6 km)	0	16 min	35

*Estimated maximal oxygen intake in ml · kg⁻¹ · min⁻¹

From B.J. Sharkey, 1997, *Fitness and work capacity*, 2nd ed. (Missoula, MT: USDA Forest Service).

PART III
Employee Selection Practices

© AP Photo/The Paris News, G.J. McCarthy

**Obstacles are those frightful things you see
when you take your eyes off the goal.**

Hannah More

We've done a job task analysis and developed and validated a job-related test. Now it is time to implement the test in the workplace. Chapter 7 begins with the least difficult step, implementing the test for new recruits. The test, by itself, is an important step toward a safer, healthier, and more productive workforce. It reveals who can and who cannot do the job and who is most likely to be injured, to have lost-time injuries, or to go out early on a disability. Consider again this quote from 1914:

> That a man must be physically sound for his work we know,
> but a standard of soundness has never been defined . . .
> it is urgent that a simple but effective method be used by all
> employing officers to ensure the rejection of the clearly unfit.

Coert Dubois, U.S. Forest Service

In a physically demanding occupation, a well-devised job-related test does not have an unfair impact on women or older applicants if applicants train for the job. And such a test will not unfairly affect racial or ethnic groups. A valid work capacity test is essentially gender-, age-, and color-blind.

A job-related test is important, but it isn't the most essential ingredient for ensuring the safety, health, and productivity of employees. The most essential ingredient is a comprehensive employee health and wellness program, combined with annual work capacity testing and job-related fitness training. Chapter 8 describes the challenges of moving from recruit testing to periodic tests for incumbents. Finally, chapter 9 provides ways to evaluate the costs and benefits of testing and training programs.

Chapter 7
Testing New Employees

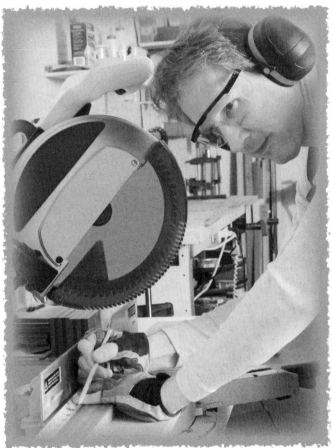

© iStockphoto.com/Eliza Snow

In 1996 the City of Erie, Pennsylvania, developed a new physical agility test as a device to screen police officer candidates. The police developed the new test without consulting any experts in physical abilities, job analysis, physical job requirements, test validation, exercise physiology, or industrial or organizational psychology. The new test consisted of a 220-yard (201 m) run with obstacles, climbing a 6-foot-high (183 cm) wall, climbing through a window opening 3 feet (91 cm) above the ground, crawling under a 2-foot-high (60.8 cm) platform, and climbing a 4-foot (121.4 cm) wall. Immediately after the run, candidates performed push-ups and sit-ups. The new test was used for several years, with several modifications in procedures.

After several years of use, the procedure was challenged, with the U.S. Department of Justice alleging that the test violated Title VII of the Civil Rights Act of 1964, that the city's use of the test had a disparate impact on female applicants, and that the test was neither related to the position of entry-level police officer nor consistent with business necessity (*United States v. City of Erie, Pennsylvania*, 2005).

The County Human Relations Commission and women's advocacy groups complained about the disparate passing rates, focusing on the 6-foot (183 cm) wall as being excessively difficult for women and not adequately job related to justify its use. The test was modified with the option to climb either the wall or a 6-foot-high (183 cm) chain-link fence. In 2004 the U.S. Department of Justice moved for a summary judgment on the issue that the test caused a disparate (adverse) impact on female candidates. The city conceded the point. In 2005 a bench trial was held to determine the job-relatedness of the test and whether the passing standard was consistent with business necessity. The judge's decision can be found in the case study at the end of the chapter.

This chapter describes the steps necessary to prepare for and administer a carefully developed job-related test of work capacity to new employees in public and private organizations.

This chapter will help you do the following:

- Identify standards in a medical examination.
- Establish an essential safety plan.
- Develop instructions for test administrators and recruits.
- Write an effective job description and recruiting advertisement.
- Schedule pretest practice and training sessions.
- Determine the number of test attempts.
- Integrate work hardening into recruit training.

Medical Standards

Applicants should receive medical clearance before training, engaging in a strenuous work capacity test, or performing the duties of the occupation. According to the Americans with Disabilities Act (ADA), an employer may not conduct a medical examination until he or she makes a conditional job offer to the applicant. Employers should conduct other parts of the application process before investing in the medical examination. The medical standards employers use should be those that are developed and validated for the occupation. In reality, however, the situation is a bit more complicated. For example, sedentary recruits should undergo health screening prior to engaging in training for the test.

The American Heart Association and the American College of Sports Medicine have published screening guidelines for those beginning a fitness program (see the Health Screening Questionnaire on page 69). The questionnaire identifies people who should see their physician before undertaking a fitness program. Applicants who check a box in section A or more than two boxes in section B should consult with their physician. The body mass index chart (BMI) can be used to determine whether a person is overweight (BMI 25 or above) (see figure 7.1). The BMI is a useful tool to determine overweight or obesity in a population, but it isn't perfect. Very muscular people (weightlifters, football players) may appear to be overweight when they are not overfat. Less emphasis should be placed on the BMI as a measure of overweight when the waist measurement is under 40 inches (102 cm) for men and under 35 inches (89 cm) for women.

Because sedentary people are 56 times more likely to experience cardiovascular problems during vigorous exertion than active people are, sedentary recruits should become active *before* beginning training. Four to six weeks of walking accomplishes the transition from sedentary to active living, which carries a much lower cardiovascular risk. Sedentary people should walk most days of the week, increasing distance and then pace, until they can walk 3 miles (4.8 km) in 45 minutes. Then, depending on their level of fitness and body weight, training for the test and the job could take six to eight additional weeks, or more. Some agencies integrate the fitness training into the recruit program. They make a conditional job offer, administer a medical examination, and begin recruit training. At the end of training, failure to pass the test could lead to conditional employment or to delayed entry, pending more physical training and successful completion of the test.

Candidates who don't pass the medical examination should not begin training, take the job-related test, or begin work. They can see their own physician to schedule additional tests or appeal the medical decision. If the candidate has a disability that would not impose a health risk or interfere with performance of job duties, the organization should consider a waiver or a reasonable accommodation. A simple example is a candidate who fails the vision test without glasses but can pass with corrective lenses. We discuss other accommodations in later chapters. It should be noted that

Assess your health needs by marking all true statements.

Section A History

You have had:

- [ ] A heart attack
- [ ] Heart surgery
- [ ] Coronary angioplasty (PTCA)
- [ ] Pacemaker/implantable cardiac defibrillator/ rhythm disturbance
- [ ] Heart valve disease
- [ ] Heart failure
- [ ] Heart transplantation
- [ ] Congenital heart disease
- [ ] Personal experience or a doctor's advice of any other physical reason that would prohibit you from carrying out the duties of the job

Symptoms

- [ ] You experience chest discomfort with exertion.
- [ ] You experience unreasonable breathlessness.
- [ ] You experience dizziness, fainting, blackouts.
- [ ] You have musculoskeletal problems (spine, knees, etc).

Other Health Issues

- [ ] You are pregnant.
- [ ] You take heart medications.

Section B Cardiovascular Risk Factors

- [ ] You are a man older than 45 years.
- [ ] You are a woman older than 55 years, you had a hysterectomy, or you are postmenopausal.
- [ ] You smoke.
- [ ] Your blood pressure is greater than 140/90, you don't know your blood pressure, or you take blood pressure medication.
- [ ] You are more than 20 pounds (9 kg) overweight (BMI > 25).
- [ ] Your blood cholesterol level is greater than 240 mg/dL, you don't know your cholesterol level, or you take cholesterol medication.
- [ ] You have a close blood relative who had a heart attack before age 55 (father or brother) or age 65 (mother or sister).
- [ ] You are a diabetic or take medicine to control your blood sugar.
- [ ] You are physically inactive (i.e., you get less than 30 minutes of physical activity at least three days per week).

Applicants who check a box in section A or more than two boxes in section B should consult with their physician.

Adapted from AHA/ACSM, 1998, "AHA/ACSM joint statement: Recommendations for cardiovascular screening, staffing, and emergency policies at health/ fitness facilities," *Medicine & Science in Sports & Exercise* 30(6): 1009-1018.

© Human Kinetics

Sedentary people should walk most days of the week, increasing distance and then pace until they can walk 3 miles (4.8 km) in 45 minutes.

Weight (lb) \ Height (in.)	49	51	53	55	57	59	61	63	65	67	69	71	73	75	77	79	81	83
66	19	18	16	15	14	13	12	12	11	10	10	9	9	8	8	8	7	7
70	20	19	18	16	15	14	13	13	12	11	10	10	9	9	8	8	8	7
75	22	20	19	17	16	15	14	13	12	12	11	10	10	9	9	9	8	8
79	23	21	20	18	17	16	15	14	13	12	12	11	11	10	9	9	9	8
84	24	22	21	19	18	17	16	15	14	13	12	12	11	11	10	10	9	9
88	26	24	22	20	19	18	17	16	15	14	13	12	12	11	11	10	10	9
92	27	25	23	21	20	19	17	16	15	15	14	13	12	12	11	11	10	10
97	28	26	24	22	21	20	18	17	16	15	14	14	13	12	12	11	10	10
101	29	27	25	23	22	20	19	18	17	16	15	14	13	13	12	12	11	10
106	31	28	26	24	23	21	20	19	18	17	16	15	14	13	13	12	11	11
110	32	30	27	26	24	22	21	20	18	17	16	15	15	14	13	13	11	11
114	33	31	29	27	25	23	22	20	19	18	17	16	15	14	14	13	12	12
119	35	32	30	28	26	24	22	21	20	19	18	17	16	15	14	14	13	12
123	36	33	31	29	27	25	23	22	21	19	18	17	16	16	15	14	13	13
128	37	34	32	30	28	26	24	23	21	20	19	18	17	16	15	15	14	13
132	38	36	33	31	29	27	25	23	22	21	20	19	18	17	16	15	14	14
136	40	37	34	32	29	28	26	24	23	21	20	19	18	17	16	16	15	14
141	41	38	35	33	30	28	27	25	24	22	21	20	19	18	17	16	15	15
145	42	39	36	34	31	29	27	26	24	23	22	20	19	18	17	17	16	15
150	44	40	37	35	32	30	28	27	25	24	22	21	20	19	18	17	16	15
154	45	41	38	36	33	31	29	27	26	24	23	22	20	19	18	18	17	16
158	46	43	40	37	34	32	30	28	26	25	24	22	21	20	19	18	17	16
163	47	44	41	38	35	33	31	29	27	26	24	23	22	20	19	19	18	17
167	49	45	42	39	36	34	32	30	28	26	25	23	22	21	20	19	18	17
172	50	46	43	40	37	35	32	30	29	27	25	24	23	22	21	20	19	18
176	51	47	44	41	38	36	33	31	29	28	26	25	23	22	21	20	19	18
180	52	49	45	42	39	36	34	32	30	28	27	25	24	23	22	21	20	19
185	54	50	46	43	40	37	35	33	31	29	27	26	25	23	22	21	20	19
189	55	51	47	44	41	38	36	34	32	30	28	27	25	24	23	22	20	20
194	56	52	48	45	42	39	37	34	32	30	29	27	26	24	23	22	21	20
198	58	53	49	46	43	40	37	35	33	31	29	28	26	25	24	23	21	20
202	59	54	50	47	44	41	38	36	34	32	30	28	27	25	24	23	22	21
207	60	56	52	48	45	42	39	37	35	33	31	29	27	26	25	24	22	21
211	61	57	53	49	46	43	40	38	35	33	31	30	28	27	25	24	23	22
216	63	58	54	50	47	44	41	38	36	34	32	30	29	27	26	25	23	22
220	64	59	55	51	48	44	42	39	37	35	33	31	29	28	26	25	24	23
224	65	60	56	52	49	45	42	40	37	35	33	31	30	28	27	26	24	23
229	67	62	57	53	49	46	43	41	38	36	34	32	30	29	27	26	25	24
233	68	63	58	54	50	47	44	41	39	37	35	33	31	29	28	27	25	24
238	69	64	59	55	51	48	45	42	40	37	35	33	32	30	28	27	26	24
242	70	65	60	56	52	49	46	43	40	38	36	34	32	30	29	28	26	25
246	72	66	61	57	53	50	47	44	41	39	37	35	33	31	29	28	27	25
251	73	67	63	58	54	51	47	45	42	39	37	35	33	32	30	29	27	26
255	74	69	64	59	55	52	48	45	43	40	38	36	34	32	31	29	28	26
260	76	70	65	60	56	52	49	46	43	41	39	36	34	33	31	30	28	27
264	77	71	66	61	57	53	50	47	44	42	39	37	35	33	32	30	29	27
268	78	72	67	62	58	54	51	48	45	42	40	38	36	34	32	31	29	28
273	79	73	68	63	59	55	52	48	46	43	40	38	36	34	33	31	30	28
277	81	75	69	64	60	56	52	49	46	44	41	39	37	35	33	32	30	29
282	82	76	70	65	61	57	53	50	47	44	42	40	37	35	34	32	30	29
286	83	77	71	66	62	58	54	51	48	45	42	40	38	36	34	33	31	29
290	84	78	72	67	63	59	55	52	48	46	43	41	39	37	35	33	31	30
295	86	79	74	68	64	60	56	52	49	46	44	41	39	37	35	34	32	30
299	87	80	75	69	65	60	57	53	50	47	44	42	40	38	36	34	32	31
304	88	82	76	70	66	61	57	54	51	48	45	43	40	38	36	35	33	31
308	90	83	77	71	67	62	58	55	51	48	46	43	41	39	37	35	33	32
312	91	84	78	72	68	63	59	55	52	49	46	44	41	39	37	36	34	32

Legend:
- ☐ Underweight (<19)
- Desirable (19–24)
- Increased health risks (25–29)
- Obese (30–39)
- Extremely obese (>40)

Figure 7.1 Body mass index chart.

Reprinted, by permission, from B.J. Sharkey and S.E. Gaskill, 2007, *Fitness and health*, 6th ed. (Champaign, IL: Human Kinetics), 242.

During the great U.S. forest fires of 1910, a frazzled ranger called the regional office to request more firefighters. When asked how many he needed, the ranger replied: "Send me ten men if they wear hats, and if they wear caps, I'll need thirty." The distinction was that the respectable, hardworking lumberjack always wore a felt hat, whereas the pool hall boys and general stew bums usually wore caps and shoved their hands deep in their pockets (Pyne 2001).

more serious medical problems are often difficult to accommodate in public safety occupations such as firefighting and law enforcement.

Safety Considerations

A department administering a job-related test of work capacity for firefighters needs to conduct a job hazard analysis and create an emergency plan for test administration.

A job hazard analysis considers the hazards or risks involved in training and testing and outlines the steps necessary to eliminate or minimize the risks. For example, it may be important to have a paramedic or emergency medical technician (EMT) and emergency equipment available during the administration of a strenuous test. A job hazard analysis includes the hazards associated with exercise training and testing, equipment, and facilities; anticipates environmental risks; and identifies the training and preparation required for personnel (e.g., CPR for instructors). The form below provides an example of the job hazard analysis for wildland firefighters.

Involving current employees in the development of the job hazard analysis is important because they understand the job best and have the hands-on knowledge that is critical to identifying potential hazards. A thorough job hazard analysis also is likely to result in fewer worker injuries and illnesses; promote safer, more effective work methods; reduce workers' compensation costs; and increase worker productivity.

An emergency plan identifies phone locations, emergency numbers, first responders, emergency equipment, and evacuation plans. For strenuous tests, an emergency medical technician or paramedic and ambulance should be available. If older employees are involved, a defibrillator may be required. All people involved in training and testing must be appropriately certified and aware of the plan. Both the job hazard analysis and the emergency plan should be in writing, signed by the appropriate people, and posted in the training and testing areas.

Job Hazard Analysis

Job title: <u>Wildland firefighter</u>   Date: _____

Tasks/procedures	Hazards	Abatement actions
Hiking	Slips, falls	Equipment, boots Qualifications and experience Training: Hike/work on hills
Hand tool use	Cuts, blisters	Safety guards, gloves
Chainsawing	Cuts, falling trees Hearing	Training, chaps, spotter Hearing protection
Smoke exposure	Particulate, carbon monoxide	Tactics: Flanking, etc.

This is a brief sample of an analysis that could include several pages of detailed abatement procedures.

The job hazard analysis anticipates environmental risks to workers such as exposure to smoke or chemicals.

Instructions

Providing detailed and clear information to test administrators and the recruits taking the test decreases the potential perception of unfairness. If testing takes place in more than one location, the test administrators must follow the instructions carefully to ensure an objective evaluation. Instructions should cover test basics, timing, rest intervals, hydration, and things that are prohibited or should be avoided. For example, the wildland firefighter pack test does not permit jogging or running. Those who continue to jog or run after a warning are disqualified. Because the test is administered in hundreds of locations throughout the country, an instruction booklet and video have been prepared to train administrators. The instructions describe how to lay out a course, assistance required during the test, and safety issues. They also include weighing the pack, pack adjustments, footwear, and other details. A table of altitude adjustments was developed to adjust tests administered above 4,000 feet (1,219 m) of elevation.

Instructions for test takers include a health screening questionnaire, medical requirements, training tips, clothing, pacing, and hydration. Candidates are told that they will be stopped if they fall too far behind the pace required to pass. They are encouraged to stop if they don't feel well and are told that they can reschedule the test another day. Each year 25,000 candidates qualify for firefighting positions with U.S. federal agencies.

Recruitment

Employers need to write effective job descriptions and recruiting advertisements that specify job demands, the work capacity test, and other requirements. Most refer applicants to a Web page for forms and additional information. Many also send out a pamphlet that describes education, training, and other requirements. Information about medical standards, the job-related test, and how to prepare for the test could be included.

What happens when too many applicants apply for the job? When confronted with 4,500 applicants for 45 positions, the City of St. Paul Fire Department adopted a tiered approach. They set up a mentoring program and held orientation sessions for the physical abilities test. Some applicants found the expectation of moving a 175-pound (79.5 kg) dummy daunting and deselected themselves from the test. A 1.5-mile (2.4 km) run with a pass time of 11 minutes 40 seconds was used to cull those with little likelihood of success.

One purpose of a test of work capacity is to reduce the field to a manageable number. New York City has over 20,000 applicants for jobs in the fire department. Lotteries have even been used successfully because employers can't hire all the "qualified" applicants. The goal of hiring the best qualified applicants can be accomplished by hiring those with the top scores. Hiring top-ranked applicants eliminates subjectivity and the nepotism that once existed in public safety occupations.

Pretest Training

As part of an affirmative action/equal opportunity program, many departments and agencies have instituted pretest training programs designed to help candidates meet work capacity standards. If participants have been inactive, they should begin with a transition to activity. When muscular fitness is important, the program should emphasize job-related strength and muscular endurance training (see case study).

A more intense training program may achieve similar results in less time. Strength typically improves 1 to 3 percent per week, with previously untrained people increasing at a faster rate. With hard training, motivated people may temporarily achieve a rate of 4 to 5 percent improvement per week. The rate of improvement will decrease or plateau as the person approaches maximal strength. In some studies, the rate of improvement in strength may be slowed if strength and strenuous aerobic training are combined (Bell et al. 2000; Hickson 1980). Strength training is specific; improvements take place in the muscles trained. A previously sedentary person on an adequate diet can expect to increase strength 50 percent or more in six months (Sharkey and Gaskill 2006).

Muscle endurance usually increases 10 percent per week when training with 15 to 25 repetitions maximum (RM). Muscular endurance has been found more effective than strength training for endurance performances (Washburn et al. 1982). With aerobic training, young adults can improve 20 to 25 percent or more, depending on their initial level of fitness. These improvements in strength, muscular endurance, and aerobic fitness are sufficient to minimize the likelihood of experiencing adverse impact on a job-related test.

Borderline candidates, those who arrive unfit and untrained and barely pass the minimum qualifications for employment, rarely get better and often become chronic problems for the duration of their employment. A well-devised entry test should reduce the likelihood of hiring borderline candidates.

Pretest training should focus on the physical demands of the job and include core training. If

CASE STUDY

Material Handling in the Military

Knapik (1997) examined the influence of a fitness training program on militarily relevant manual material handling capability (loading, unloading, and moving goods). Thirteen women trained for 14 weeks, performing resistance training three days per week and running with interval training two days per week. Subjects increased maximal lifts by 19 and 16 percent for floor-to-knuckle height and floor-to-chest height, respectively. They improved 10-minute lifting endurance with 33 pounds (15 kg) from 167 to 195 lifts (16.7 percent), with no change in perceived exertion during the work. Body fat decreased (9 percent), and fat-free mass increased (6 percent). The generalized fitness training program, conducted for about one hour per day, substantially improved women's manual material handling ability.

A modified basic training program that included resistance training was compared with standard basic training for its effects on material handling performance of army recruits (Williams, Rayson, and Jones 2002). Forty-three males and nine females performed the modified training for 11 weeks. Significant improvements with the modified training occurred in all six material handling tests and the loaded march with 15 kilograms (33 lb). Significantly greater improvements were observed following the modified training compared with the standard training in maximal box lift, dynamic lift to 1.45 m, loaded march performance, and $\dot{V}O_2$max. The authors concluded that improvements in material handling and other aspects of fitness can be enhanced by using a training program that includes carefully designed resistance training. Of particular note were the strength-related material handling tasks that are encountered in the military.

strength is essential, demanding aerobic training should not be included unless it is needed on the job. If muscular strength is the limiting factor, candidates should train for strength. If strength and muscular endurance are essential, pretest training should include both facets of muscular fitness (see table 7.1).

Table 7.1   Muscular Fitness Training

	STRENGTH	MUSCULAR ENDURANCE
For	**Maximum force**	**Persistence with load**
Prescription	6-10RM* 3 sets 3 days/wk	15-25RM** 3 sets 3 days/wk

*RM = repetitions maximum, the most you can do with weight

**If power is needed, use 30-60% of max strength and work as fast as possible.

Pretest training should focus on the job's physical requirements. If muscular strength is the limiting factor, trainees should train for strength.

©Photodisc

To achieve muscular fitness, trainees must train three days per week. For strength, they should do one set with 10RM (repetitions maximum), one with 8RM, and one with 6RM. After training for four to six weeks, trainees should add muscular endurance training with one set of 15RM. As strength becomes adequate, they should do all three sets for muscular endurance.

Retesting

How many opportunities should a recruit have to pass the test, and how much time should pass between trials? It makes sense to provide at least one additional opportunity to pass the test, if only to avoid grievances. Those who don't pass the second test should be encouraged to train before they retake the test, unless they were very close to passing. Prior to testing, employers should publish a schedule of test dates, with the policy on retests. Training for several weeks will allow many to pass on the second trial. When time allows, one or two additional trials are acceptable. Those who fail three times should have to wait three to six months before their next attempt, to allow time for training adaptations. When there is not sufficient time for a retest, as with the emergency hiring of wildland firefighters, a single test is justified.

Work Hardening

After passing the job-related test, recruits will need additional preparation. Work hardening is a gradual progression of work-specific activities designed to bring trainees to the job ready to deliver a good day's work. Although fitness training provides the foundation, it is no substitute for job-specific work hardening. Aerobic and muscular fitness training increase the strength of tendons, ligaments, and connective tissue. Work hardening ensures that the muscles and connective tissue used on the job are tough and ready to go. Feet are work hardened when recruits train in the boots they will use on the job. There is no substitute for preparing the trunk and upper-body muscles for prolonged work with hand tools. Using the tools used on the job prepares the back and shoulder muscles for prolonged work. This work also toughens the hands so workers won't get blisters the first day on the job (Sharkey 1997b).

Work hardening ensures that the muscles and connective tissue a worker uses on the job are tough and ready to go.

CASE STUDY

Military Fitness

Few recruits enter U.S. military service with the physical fitness they need for the arduous duties they will confront. All the branches have fitness standards for active-duty personnel, and all administer physical fitness tests during and after basic training. But none of the military branches currently tests its recruits prior to basic training, resulting in some recruits with low levels of fitness and an elevated risk for injury and attrition. Testing prior to basic training could identify high-risk recruits who would benefit from taking part in prebasic fitness training prior to entry into basic training. Recruits who fall too far below standards could have their entry into the service denied or delayed until they meet entry standards.

Cardiorespiratory fitness tests for the U.S. military services are as follows: air force and navy, 1.5-mile (2.4 km) run; army, 2-mile (3.2 km) run; and marine corps, 3-mile (4.8 km) run. Concerned with the body size bias in running tests, researchers developed the backpack run test, a model for a fair and occupationally relevant military fitness test (Vanderburgh and Flanagan 2000). The purpose of the study, published in the journal *Military Medicine*, was to develop and validate a test consisting of a 2-mile (3.2 km) run wearing a 66-pound (30 kg) pack. Fifty-nine male service academy cadets performed an unloaded 2-mile (3.2 km) run. The researchers calculated the oxygen cost of the unloaded run and the cost of the 66-pound backpack run. Correlations between body weight and the loaded backpack run demonstrated that the bias against heavier runners in cardiorespiratory (aerobic) tests is eliminated with the backpack run. The test requires standard issue equipment, demonstrates occupational and health-related components of fitness, and eliminates the body size bias of unloaded tests. The authors recommended that the military consider the procedure when developing or modifying tests of physical fitness.

Summary

This chapter outlined the steps necessary to prepare for and administer a job-related test of work capacity. We reviewed medical standards, the job hazard analysis, and the emergency plan. We discussed the importance of test administrators providing clear instructions to recruits taking the test. And we discussed retest opportunities and the need to schedule retests.

An important part of the chapter dealt with pretest practice and training sessions. Pretest training sessions should be designed to improve test and job performance. Training three times per week for eight weeks will achieve significant results, although some may require more time to make adjustments. Integrating pretest training into recruit training will ensure participation in the training program. However, it involves paying recruits who might not pass the test at the end of training. If you think pretest training sounds expensive, keep in mind that it is a bargain compared to the costs of mounting a legal defense against a complaint of disparate or adverse impact.

We also discussed the need to integrate work hardening into recruit training so new employees come to the job hardened and ready to work. To see how selectivity in hiring affects work output, see appendix D on page 199.

CASE STUDY

United States v. City of Erie, Pennsylvania (2005)

The United States District Court for the Western District of Pennsylvania did not question the City of Erie's good faith in developing and administering a physical agility test as a device to screen police officer candidates. The city made genuine efforts to develop a test that would meet the bureau's needs, while faced with budgetary constraints and objections raised by the police officers' union. The court noted that the city took steps to make hiring more amenable to females by modifying its administration over the years. The court was sympathetic to the police officers' interest in maintaining the physical integrity of the force and did not advocate watering down hiring standards in such a fashion as to jeopardize the lives of the officers or the health and safety of the public. Finally, the court noted that nothing in the business necessity standard requires employers to hire employees in numbers reflecting the ethnic, racial, or gender makeup of the community.

The court ruled that the test development process did not establish the job-relatedness of the procedure and that failure to establish job-relatedness made it difficult to support the passing standard. Finally, the court ruled that the 90-second cutoff score was not consistent with business necessity in that the city had not proved by a preponderance of evidence that the entry-level standard corresponded to the minimum qualifications necessary to successfully perform the job of police officer. Finally, consider this passage from the trial transcript for the Erie case:

"During the administration of the 1998 PAT, an 11-year-old girl, described as 'petite,' 'wiry,' 'very active' and a 'gymnast,' observed the test and requested to take it unofficially. She passed all elements within the allotted 90 seconds."

Chapter 8

Testing Incumbent Employees

© Getty Images/Mel Svenson

Studies of fire and law enforcement personnel indicate the dismal level of physical readiness of many emergency service personnel. Fitness declines rapidly in this sedentary population, and body fat levels double during a typical 20-year career (Davis and Starck 1980). With this physical decline comes a high rate of heart disease, which, surprisingly, is viewed as a job-related disability. Taxpayers deserve the best public servants available, and incumbents deserve to remain healthy and fit for the job. The solution is to test recruits and follow up with an annual performance evaluation, coupled with a job-related physical maintenance (fitness) program.

This chapter examines the process of negotiating and implementing a job-related testing program for incumbent employees in physically demanding occupations.

This chapter will help you do the following:

- Discuss the relationship between age and job performance.
- Understand the concept of physiological age.
- Understand the necessity of periodic testing.
- Negotiate periodic testing and implementation options.
- Provide adequate notice of testing as well as job-related fitness training.
- Ensure safety and medical surveillance.
- Identify retest options and consequences of failure.

Age and Performance

Aerobic fitness peaks in the 20s and then begins to decline. The rate of decline is influenced by physical activity and training (see figure 8.1). Although the sedentary person declines at 9 to 10 percent per decade (1 percent per year), being active cuts that rate in half (4 to 5 percent per decade), and training can cut the active rate in half (2 percent per decade) (Kasch et al. 1988), for 25 to 33 years (Kasch 2001). On the positive side, aerobic training

CASE STUDY
Exemption to the ADEA

The Age Discrimination in Employment Act (ADEA) of 1967 outlawed job discrimination based on age, except when age was a bona fide occupational qualification (BFOQ). The BFOQ requires that the effects of advancing age make it impossible for all or most people above a certain age to do a job, or it is impossible to assess work capacity with a job-related test, or both.

As you will see, neither condition of the BFOQ is defensible. And yet the U.S. Congress exempted fire and law enforcement personnel from the ADEA, assuming that age was a BFOQ. Because of the exemption, most departments have instituted a maximal hiring age (e.g., 35) or a mandatory retirement age (e.g., 55 or 60). When it came time to review the exemption, the Equal Employment Opportunity Commission (EEOC) funded a large study to evaluate the effects of age on performance. The study found that age was poorly correlated with job performance and that job-related tests accurately reflected work capacity (Landy 1992). Unfortunately, Congress ignored the comprehensive study, listened to representatives of labor unions, and voted to reinstate the exemption to the ADEA.

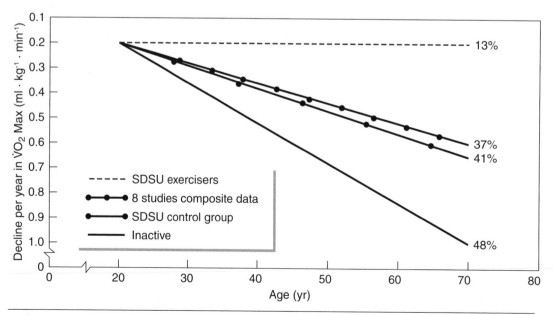

Figure 8.1 Decline in maximal oxygen intake with age (SDSU = San Diego State University).

Adapted from F. Kasch et al., 1988, "A longitudinal study of cardiovascular stability in active men aged 45-65 years," *Physician and Sportsmedicine* 16(1): 117-126.

can increase aerobic fitness 25 percent (and more with weight loss).

The American College of Sports Medicine said: "Age in itself does not appear to be a deterrent to endurance training" (1998); improvements occur regardless of age, even for the elderly. The rate of decline in $\dot{V}O_2$max after 50 years of age in endurance-trained people was related to a decline in training volume (Pimentel, Gentile, and Tanaka 2003). Those who increase their level of activity and maintain the increase can slow the rate of decline until the sixth or seventh decade of life. Of course, those who do not train or engage in regular physical activity face an inexorable decline, at the rate of 1 percent per year.

Strength declines rather slowly with age until the fifth decade, when the rate of decline increases. This loss of muscle mass has been called sarcopenia, or vanishing flesh. Sarcopenia is an important component of frailty in the elderly, contributing to a loss of strength, increased falls, and fractures (Welle and Thornton 2002). Sarcopenia results from a loss of muscle fibers and fiber atrophy,

because of lack of use, and decreased muscle-building hormones (testosterone). People who use their strength regularly retain muscle function much longer. Also, recent evidence shows that very elderly men and women (average age 87 years) can counter muscle weakness and frailty with resistance training (Fiatarone et al. 1994).

Muscular fitness declines at a rate that is inversely related to use. Those who use their muscles regularly maintain strength into their 50s and beyond, whereas those who fail to use their muscles experience a more rapid decline.

Physiological Age

Physiological age (also called functional age) is a composite of health, physiological capacity, and performance measures. The best single measure of physiological age is probably the aerobic fitness score. It addresses the health and capacity of the respiratory, circulatory, and muscular systems. Moreover, a considerable body of evidence shows an inverse relationship between aerobic fitness and a number of risk factors. Thus, it is possible at age 55 to have the health and performance capabilities of the average 25- to 30-year-old. This fact has considerable relevance when it comes to society's view of aging and its consequences, such as age discrimination in hiring. Aging does not ensure a

© Human Kinetics

People who strength train regularly can retain muscle function for much longer than those who do not.

ⅠⅠⅠⅠⅠⅠⅠ **TOO OLD TO JUMP?** ⅠⅠⅠⅠⅠⅠⅠ

For many years managers considered George Cross too old to become a smoke jumper, but not too old to conduct the physical training program for this elite bunch of wildland firefighters. Smoke jumpers parachute into the wilderness to battle wildfires. When the fire is contained, they load up their gear, all 125 pounds (56.7 kg), and pack it out to the nearest road. As he approached retirement from his university teaching position, George decided to sign up for jumper training, at the tender age of 58. The agency had conducted a study of age and performance and decided to remove maximal hiring and retirement ages (Sharkey 1987). They found that physiological age, not chronological age, was associated with performance. George passed the arduous physical tests, jump training, and field tests such as the pack-out, and he became a smoke jumper.

rapid decline in performance, and when physical performance is important, the physiological age is a more accurate predictor of performance potential than chronological age (Sharkey 1987). Other performance-related indicators of physiological age include muscular fitness, body composition, and health habits. The only way to determine the capacity to perform in physically demanding occupations is to conduct periodic evaluations of work capacity.

Periodic Testing

Many programs use job-related tests for recruits, but few require them for incumbents. Why don't more departments require regular tests to ensure the maintenance of work capacity? Reasons range from tradition and lack of information to fear of failure. Few jobs allow new employees to coast an entire career until retirement. Law enforcement personnel are tested to be sure they are maintaining firearms skills. Paramedics are constantly upgrading skills and knowledge. And we all have had to learn and upgrade our computer and electronics skills as technology evolves. Yet employees and their unions in many professions have steadfastly refused to submit to the documentation of the maintenance of work-related physical capabilities. This isn't true in all demanding occupations.

The U.S. military maintains fitness requirements, even in the highest ranks. Wildland firefighters take job-related tests before each fire season. The elite national fire crews, the Hotshots, are 20-person crews that are ready to travel to the next big fire, where they will work long shifts until the fire is controlled. They are not satisfied with their fitness standards. They have requested higher standards that reflect their level of fitness and work capacity. Elite units, such as U.S. Army Rangers, U.S. Navy SEALs, military special forces, smoke jumpers, and Hotshots, are motivated by high standards. One way to improve the performance of a group is to raise entry and job-related fitness standards. Incumbents will improve, and organizations will attract quality applicants.

Initiate Dialogue

A program of periodic testing should begin with an initial dialogue and negotiations of implementation options. There are several ways to imple-

⁞⁞⁞⁞⁞ IMPLEMENTING THE ⁞⁞⁞⁞⁞ FIVE-MINUTE STEP TEST

In 1975, after several years of field study and approval by the Civil Service Commission, managers of five U.S. federal land management agencies decided to implement an annual test for wildland firefighters. The five-minute step test predicted maximal oxygen intake ($\dot{V}O_2$max) and had a passing score of 45 (ml · kg^{-1} · min^{-1}), regardless of age or gender. In spite of efforts to explain the test and its purpose, management received a number of questions and concerns regarding the test. Managers arranged for the test developers to visit a number of field locations and discuss the procedure with employees. The developers assembled a slide presentation and packaged it with an audiotape for use around the country. They also developed training materials for wildland firefighters (Sharkey 1977). In time the test was accepted as an indicator of work capacity. The test remained in use until 1998, when it was replaced with the job-related pack test.

ment mandatory tests for incumbents. One way is for management to mandate the requirement. This top-down approach is likely to cause concerns among incumbents and could lead to the filing of grievances. And yet it has been done (see the box above, Implementing the Five-Minute Step Test).

Because the top-down approach is unlikely to be successful in today's work climate, it is essential to initiate a dialogue between management and labor to discuss, negotiate, and plan an acceptable implementation. Sometimes the impetus comes from above: Management says, "We will have this program, so let's discuss how we are going to get there." Sometimes the motivation comes from the workforce, as was the case when the Hotshot firefighters sought higher standards. And sometimes tests of incumbent employees are the outcome of a sincere discussion of how to maintain or improve performance, safety, and health. Just keep in mind the old quote, "If you don't know where you are going, you may end up someplace else."

Employers hoping to implement a test of incumbent employees should begin a dialogue with the employees. If employees were involved in the job task analysis and test development, they already know a lot about the test. After early discussions, it may help to invite in an acknowledged expert, one with years of experience working with the occupational group. The expert can describe individual and organizational benefits and how to achieve them. Employers should listen to comments and concerns from employees and take notes. In the next section we indicate ways to smooth the path to implementation.

Negotiate Implementation Options

In many settings test implementation involves discussions with members of the union. At some point it is time to list concerns and how they can be mitigated. Part of this process is to identify benefits and strategies for implementation. Benefits can range from salary incentives to benefit enhancements, such as reductions in individual contributions to health insurance premiums. Strategies for implementation range from a gradual start-up to immediate implementation for all incumbent employees. We will not deal with financial incentives, but we will look at strategies.

Phase-In

The phase-in approach can begin with new recruits, saving annual tests for incumbents until the current year's recruits have been tested annually for one or more years. The phase-in could also use incentives to bring incumbents into the program.

Grandfathering

Some organizations elect to "grandfather" older employees. This means that older employees are exempt from the test program requirements. Although this approach may ease the negotiations, it does not look good in a legal challenge. It is hard to argue for the business necessity of a standard when some, who do the same work, are not held to the same standard. If used, the grandfather option should be reserved for those close to retirement or for middle managers. Older workers who refuse to submit to testing or who should not be tested for medical reasons could be offered job-related fitness training, retraining, reassignment, or, after given a certain amount of time to comply, dismissal.

All Incumbents

The decision to involve all incumbents in testing may seem hard to sell to labor, but it is the least difficult option to implement and defend in court. If an untested incumbent is injured or dies on the job, the employer could be sued based on the assertion that the untested victim wasn't physically

---

CASE STUDY

Arbitration and the Hanford Fire Department

The U.S. Department of Energy subcontracts management of the Hanford atomic energy site in the State of Washington. Hanford is engaged in one of the world's largest environmental cleanup projects. At the Department of Energy's request, the subcontractor negotiated with the Hanford Fire Department to institute an annual fitness assessment for incumbent firefighters. The fire department engages in structural and wildland firefighting on the site's Columbia Basin location. With the aid of a job task analysis, a review of the literature and job functions, and the advice of consultants, they decided to adopt the validated 2-mile (3.2 km) field test used by wildland firefighters. The test predicts an aerobic fitness score of 40 (ml · kg^{-1} · min^{-1}), a level appropriate for the type of fires they encounter (grass) and the duration of the work. All parties agreed to a program including a medical evaluation, a physical fitness program, and an annual field test.

In the first year of the test, three employees submitted grievances stating that the standard is *more difficult than necessary* and that the test does not provide any accommodation for any bargaining unit member based on age, sex, or disability; therefore, the test may cause qualified union members to suffer negative employment consequences because they may wrongfully and inappropriately be determined to be unqualified for their position with the department. The grievances were scheduled for arbitration, even though no members of the department had failed the test or suffered negative employment consequences. After two postponements and months of delay, the case came before the arbitrator, who was to hear witnesses from both sides and then render a decision. Before the proceedings were called to order, union members held a caucus that continued for several hours, culminating in a written agreement between management and labor that verified and accepted the original plan.

capable of the work or was at risk. Also, a legal challenge for adverse impact is easier to defend when all incumbents are tested. The concept of one job, one standard makes this approach easiest to defend in court. All negotiations must be recorded and signed off by management and labor. Unfortunately, that doesn't mean that the union won't file a grievance (see the case study, Arbitration and the Hanford Fire Department).

Employers should negotiate with employees and agree on all aspects of the program, including the frequency of testing. Annual or semiannual testing makes sense for most occupations. Because testing takes time and costs money, the schedule should fit the demands of the occupation. A few weeks of bed rest can cause a 30 percent decline in aerobic fitness, so those who have been out of work as a result of injury or illness may be required to pass a test before returning to duty.

Next we discuss the considerations involved in initiating a testing program. A well-designed testing program should provide adequate notice, include safety provisions, require medical clearances, and offer job-related fitness training.

Providing Adequate Notice

It is essential that incumbents be provided adequate notice well in advance of testing. To avoid misunderstandings, they should be given

Fit employees should be encouraged to mentor coworkers.

a letter from management; copies of relevant documents; an opportunity to ask questions; and details about medical issues, training, testing, and the retest policy. Employers should provide a facility, certified fitness instructors, and ample time for employees to train for the test, as well as company time for training and encouragement to do additional work during time off the clock. Fit employees can mentor coworkers. A nutrition program, modified lunches and other meals, and personal advice on weight control can help overweight employees. Some distraught incumbents may benefit from counseling. Employers should do all they can to make the initial annual test as successful as possible.

Medical Standards and Safety

The physically demanding occupation needs a job-specific medical standard and an examination that reflects the demands of the work. Many organizations use a comprehensive medical examination for recruits, with follow-up exams every five years, switching to every three years at the age of 40 or 45. Programs with a fitness/wellness program can use less frequent exams, as long as employees get periodic medical tests and complete a health risk screening questionnaire annually.

As with recruit testing, employers need an up-to-date job hazard analysis and an emergency plan for training and testing. The job hazard analysis considers the risks involved in training and testing and the steps necessary to eliminate or minimize the risks. The emergency plan identifies phone locations, emergency numbers, first responders, emergency equipment, and evacuation plans. All people involved in training and testing must be appropriately certified and aware of the plan. Both the job hazard analysis and the emergency plan should be in writing, signed by the appropriate people, and posted in the training and testing areas.

Fitness Training

A "Current Comment" published by the American College of Sports Medicine states: "As athletes strive to improve their performance through effective training techniques, so too can workers benefit from optimally planned exercise training programs designed to boost occupational physical performance." Physically demanding occupations, such as those found in the armed services, emergency rescue professions, and construction and warehouse industries, require a high degree of physical fitness. Performance in these occupations can be enhanced by participation in exercise programs targeted at improving cardiorespiratory (aerobic) and muscular fitness. The program should be designed to improve performance in tasks identified in the job task analysis (American College of Sports Medicine 2002).

In the ideal situation, job-specific training is combined with health-related fitness to improve performance and enhance health through the reduction of coronary and other risk factors. Performance in a material handling job can be augmented with muscular fitness training, with considerable attention given to core training. But that training may not achieve the important health benefits of aerobic fitness, including reductions

in blood pressure, cholesterol, and excess body weight. We describe the health-related fitness program in chapter 10 and the job-related fitness program in chapter 12.

Test Results

The consequences of not passing a job-related work capacity test should be spelled out in the labor–management agreement. As part of the phased-in implementation, employers should provide opportunities for retesting, additional training options, and ample time for adaptations to training. Training for several weeks will allow many to pass on the second trial. As with recruits, those who fail three times should have to wait several months before their next attempt. If an employee has not passed within six months of the original test, it may be time to consider other alternatives.

When incumbents don't respond to training, a medical evaluation may identify contributing factors. Overweight people will need counseling, a weight loss diet, physical activity, and a behavior modification program (Sharkey and Gaskill 2007). Those with back, knee, or other orthopedic problems may need physical therapy and specific exercises. Studies indicate a relationship between work-related physical problems and psychological problems. Counseling may identify issues that compound low back, carpal tunnel, and other musculoskeletal problems. Unfortunately, no matter what the department does to help an employee, some will not be able to pass the work capacity test.

Employers may have to consider reassignment, early retirement, or dismissal for employees who don't pass the test. Smart employees who fail a job-related test may consider advancement or transfer to a less physically demanding position, such as training, fire prevention, or community relations. When dismissal is necessary, management should do what it can to

Physically demanding occupations such as those found in the warehouse industries require a high degree of physical fitness. A worker's performance can be enhanced by participating in an exercise program targeted at improving aerobic and muscular fitness.

find employment for the person elsewhere in the system. In the end, however, public service employment is not a right; it is a privilege. The job is not for everyone, and not everyone has the capacity to meet the requirements of the job.

We do not recommend the long-term use of financial incentives as a way to promote fitness. Physical fitness is a personal responsibility as well as an occupational requirement in a physically demanding occupation. Employers should use incentives to initiate participation, but they should not rely on them for the maintenance of job-related fitness. Workers should be dedicated to service, which requires a career-long commitment to fitness. This creates what psychologists call intrinsic motivation. Employees who complain that they don't have time for fitness training should remember that our last six presidents have been able to work fitness into their schedules.

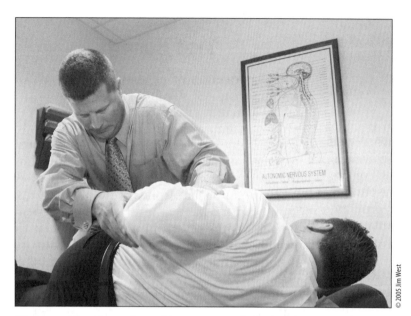

Workers with orthopedic problems may need physical therapy and specific exercises before they can take and pass a work capacity test.

© 2005 Jim West

CASE STUDY

Lanning v. SEPTA, 1999

The Southeastern Pennsylvania Transportation Authority (SEPTA) is a regional mass transit organization operating in the Philadelphia area. SEPTA's transit police officers are responsible for law enforcement along the route, including above- and below-ground stations. A job task analysis identified critical tasks, including the ability to run to apprehend and control a suspect or to provide aid to another officer. Based on the distance between stations and the need to climb up or down stairs to stations, a 1.5-mile (2.4 km) run was used to predict aerobic fitness. Additional test elements included measures of muscular fitness. The tests were applied to candidates and incumbents. In 1998 SEPTA was sued by female applicants who have been or will be denied employment by reason of their inability to meet the physical entrance requirement of running 1.5 miles (2.4 km) in 12 minutes or less. We'll say more about this fascinating case later in the book. Our purpose here is to indicate the effect of the standards on incumbents.

When incumbent testing was first introduced, officers were disciplined for failing to meet their goals. The patrol officers' union objected, claiming that the disciplinary component of testing was never the subject of collective bargaining. The union took SEPTA to arbitration over the matter and won, so SEPTA was precluded from disciplining officers who failed the incumbent testing. Management then offered an incentive program to gain compliance with fitness standards, consisting of financial rewards and health club memberships, with union concurrence.

The supervisor of patrol officers has observed the impact of the physical fitness testing program on incumbents. According to his testimony, the program has resulted in higher-caliber officers who are more vigilant in patrol and who are better able to effectuate backups and to assist fellow officers. Since the implementation of the fitness program, felony offenses (homicide, rape, robbery, aggravated assault, burglary, theft) are down by approximately 70 percent. The supervising officer believes that the fitness program has contributed to this reduction. Data collected for this trial indicate better arrest records by officers who pass the 1.5-mile (2.4 km) run test versus those who do not.

Summary

To emphasize the need for periodic work capacity testing, we began this chapter with a discussion of age and performance. Aerobic fitness declines with age, even when people remain active. The only way to ensure the maintenance of work capacity is to test employees on a regular basis. Of course, a job-related fitness program can maintain work capacity, but the only way to be certain the training is effective is to conduct regular evaluations. Voluntary programs are not effective with the majority of employees. Without annual tests or mandatory fitness programs with established benchmarks, employees in physically demanding occupations are certain to regress in fitness and gain fat at rates similar to those of the general population.

Management and labor should negotiate periodic testing and implementation options appropriate for the occupation. Options include a phase-in, an incentive approach, and beginning the program with all incumbents. Grandfathering should be reserved for unusual cases. The frequency of testing, the number of possible retests, and the consequences of failure should all be considered.

Employers should also determine in advance the options available to those who cannot pass the test. When management and employees have reached an agreement, approvals should be put in writing and the plan should be distributed to all incumbents. Prior to the initiation of testing, employees should be given adequate notice and offered job-related fitness training. Employers should provide medical examinations, conduct a job hazard analysis, and develop an emergency plan.

Initiating an annual test of work capacity is a major step in most organizations. It will be welcomed by some, be accepted by many, and cause apprehension and anxiety in a few. Employers should not be surprised if there are complaints, even a few grievances. Employees should be given time to become familiar with the test and how to prepare for it. In time employers will see improvements in fitness, performance, safety, and health. And as performance is enhanced, morale and job satisfaction often improve. Many of these factors are measurable. Chapter 9 presents ways to evaluate the impact of the test program on employee safety and health, performance, the organization, and the community.

Chapter 9
Program Evaluation

© Jim West

An employer's work is not done once it has implemented a testing or fitness program. To retain management, employee, and community support, employers must evaluate the costs and benefits of the program. Too often, evaluation is considered after the program has been in operation for some time, but it is important to plan the evaluation process before the fact to ensure that appropriate sources of data have been identified and collected. The purpose of this chapter is to demonstrate ways to evaluate the costs and benefits of job-related work capacity tests for recruits and incumbents and work-related fitness programs for employees.

The chapter will help you do the following:

- Use available information in program assessment.
- Consider the need for additional measures.
- Develop a surveillance system for injuries and medical data.
- Use appropriate analysis techniques.
- Assess and improve the cost-effectiveness of the program.
- Report the results of the evaluation.

Evaluation

Planning for a program evaluation should begin when the program is being developed. The goals and objectives of testing and fitness programs should be formulated in a collaborative effort between management and employees, with help from a qualified consultant. The next step is to decide what information should be collected and to define, in advance, the questions to be answered. The overall goal of the process is to identify ways to improve the program.

Tests of work capacity and fitness programs can be evaluated in terms of process, impact, and outcome evaluation.

Process Evaluation

A process evaluation answers questions about numbers, demographic variables, pass/fail rates, participation rates, attendance (for fitness programs), and satisfaction. Data are obtained from records and from questionnaires completed by participants. Employee health records should include a personal health history, familial health risks, and a health risk analysis. A process evaluation may be used to improve the process, instructions, or other materials.

To evaluate the process of a test, employers will need the number of workers tested and their gender, age, height, and weight, as well as race, ethnicity, religion, national origin, and disability information. Employee status (recruit or incumbent) and years with the department should be recorded and pass/fail records and test scores should be kept for all who take the test. A brief questionnaire can assess employees' feelings about the helpfulness of test instruction and the quality of test administration, as well as the time spent training for the test.

To evaluate the process of an employee fitness program, employers will need demographic, attendance, attrition, and injury information.

Impact Evaluation

An impact evaluation examines the health and behavioral changes that take place following participation in a test or fitness program. Data could include changes in aerobic and muscular fitness,

---

||||||| EVALUATION DESIGN |||||||

- Engage stakeholders: those involved, those affected, and intended users.
- Describe the program: need, expected effects, activities, resources.
- Select the evaluation design: purpose, questions, variables, methods.
- Gather credible evidence: indicators, sources.
- Use a qualitative, quantitative, or combination approach.
- Report findings or conclusions.
- Ensure use and share lessons: feedback, follow-up, dissemination.

Adapted from the Centers for Disease Control and Prevention (CDC) 1999.

CASE STUDY
Fitness for Firefighting

According to researchers (Peate, Lundergan, and Johnson 2002), firefighters work at maximal levels of exertion. Fitness for the job requires an adequate aerobic capacity. An adequate level of aerobic fitness, or $\dot{V}O_2$max, can improve a worker's ability to perform the job and reduce the risks of cardiopulmonary conditions. The authors noted that inactive firefighters have a 90 percent greater risk of myocardial infarction than those who are aerobically fit. A sample of 101 firefighters completed a questionnaire that had them gauge their fitness by responding to comments such as, "I avoid walking or exertion; always use the elevator, drive whenever possible" (low fitness), or "I run over 10 miles per week or spend 3 hours per week in comparable physical activity" (high fitness). Participants completed two measures of aerobic fitness: a step test and a submaximal treadmill test. There was no association between the firefighters' self-perception of fitness and their aerobic capacity. The authors concluded that because of the critical job demands of firefighting and the serious consequences of inadequate fitness, periodic aerobic capacity testing and individualized fitness prescriptions may be advisable for all active-duty firefighters.

cardiovascular risk factors (cholesterol, blood pressure, body weight, blood glucose, smoking), and frequency of injuries. This information may be used to support or modify the test or program.

© Hisham F. Ibrahim/Photodisc/Getty Images

An impact evaluation examines the health and behavioral changes that take place following participation in a test or fitness program. Data could include cardiovascular and other risk factors such as smoking, seen here, or drinking, which may be used to support or to modify the program.

An impact evaluation for a test includes measures related to the test and performance. An impact evaluation of a fitness program includes job-related measures of aerobic and muscular fitness, body composition (BMI, lean body weight), flexibility, and heart disease risk factors (including lipid profile with HDL and LDL cholesterol, triglycerides, and C-reactive protein). Employers should keep records of employees' use of the health care system, as well as tobacco, alcohol, and drug use and safety behaviors (e.g., the use of seat belts). A questionnaire could explore changes in physical activity, eating behaviors, stress, mood states, and job satisfaction. Psychosocial variables such as these have been associated with lost-time injuries in the workplace (see the case study below). Improved fitness could improve both areas.

CASE STUDY

Psychosocial Factors and Disability

A cross-sectional study of over 8,000 workers (4,230 men and 4,043 women) found relationships between work stressors and musculoskeletal problems, providing evidence that both physical and psychosocial factors affect disability and were affected by disability (Cole et al. 2001). Low social support at work and high job insecurity were independent predictors of restricted activity due to musculoskeletal problems. In the construction industry, working in awkward positions was associated with location-specific musculoskeletal disorders. Some psychological factors, primarily those reflecting characteristics of individual people, are strongly associated with symptoms in all body locations. Symptoms were not associated with support from workers or supervisors or control of the work situation. All location-specific musculoskeletal disorders increased with age (Engholm and Holmstrom 2006).

A study of upper-body musculoskeletal problems (neck, wrist, hand) found that wrist and hand complaints were associated with smoking habits, fewer hobbies, fewer work breaks, heavy lifting, and some personality factors (introversion) and psychosocial factors (low work appreciation). Neck complaints were also associated with physical and psychosocial factors (Malchaire et al. 2002). A study of lost time as a result of occupational low back pain in soldiers concluded that interventions should target ergonomic, workplace, and individual psychosocial risk factors (Feuerstein et al. 2001). A recent study suggested that strenuous leisure-time physical activity might play a role in the prevention of future psychological complaints, poor general health, and long-term absenteeism in a working population (Bernaards et al. 2006).

Outcome Evaluation

An outcome evaluation examines the changes that take place in the organization as a result of the test or fitness program. It considers the cost of tests and programs in relation to their benefits. Data are obtained on productivity, absenteeism, attrition, employee morale, training, workers' compensation costs, and the use of health care. This information is needed to determine the cost-effectiveness of the program (Cox 2003).

For a job-related test, an outcome evaluation could consider performance before and after testing and evaluate the relationship between test scores and measured performances. Absentee and attrition numbers could be related to employees' scores. The loss of marginal employees increases recruitment and training costs. Workers' compensation costs also burden physically demanding occupations. Tests and training programs could reduce these costs as well as the cost of health care, which has increased at double-digit rates. Regular participation in physical activity can reduce the risk of coronary artery disease by 30 to 70 percent, depending on the degree of involvement. Fit workers experience fewer lost-time injuries, miss fewer days, and recover faster from injuries than less fit workers. A program that gradually increases workers' involvement in physically demanding tasks after an injury or illness can save employers from the high costs of replacing workers and retraining their replacements.

Surveillance System

As part of the employee health program (see chapter 10), employers should keep a close watch (surveillance) on medical risks and injuries. Medical surveillance focuses on risks associated with family history (heart disease, cancer), elevated measures (cholesterol, blood pressure), and age-related risks (glaucoma, prostate-specific antigen). Each employee should be aware of the risks and the need to participate in periodic tests.

An injury surveillance system records all lost-time injuries and uses the data to determine injury trends and the need for personal protective equipment. When new protective equipment is introduced, the surveillance system will indicate its effects on injuries. The data can indicate the need for additional training, whereas the system will indicate whether the training reduced the risk of injury. Injury data can be correlated to levels of performance on the job-related test or to levels of fitness. Of course, this database must be kept confidential.

CASE STUDY

Work Performance at SEPTA

In the *SEPTA* case (see chapter 8), the plaintiffs argued that the 1.5-mile (2.4 km) run would deny employment to applicants unable to meet the requirement. SEPTA presented evidence of the value of the test standard to officer performance. When officers of similar age, tenure, and rank were compared, those who achieved the 1.5-mile (2.4 km) standard (42 ml · kg^{-1} · min^{-1}) or higher had a statistically higher number of arrests and arrest rates for all crimes than those who never met the standard. The data showed a 14 percent advantage in overall arrest rate, and a 32 percent advantage in arrests for more serious crimes. The analysis also found that street patrol officers who met or exceeded the fitness standard of 42 (ml · kg^{-1} · min^{-1}) received more commendations. In a group of 207 commendations, 96 percent went to officers who met or exceeded the standard (average $\dot{V}O_2$max = 46 ml · kg^{-1} · min^{-1}). Of 198 commendations involving arrest, 116 referred to a foot pursuit, use of force, or other physical exertion. The aerobic standard was viewed as job related and consistent with business necessity (*Lanning v. SEPTA*, 1998).

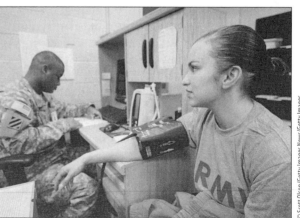

As part of the employee health program, employers should keep a close watch on medical risks such as elevated cholesterol and blood pressure.

evaluation planning process. More sophisticated statistical techniques are seldom needed for an evaluation.

INJURY SURVEILLANCE AT THE NCAA

A number of years ago, the National Collegiate Athletic Association (NCAA) began developing an injury surveillance system designed to track the incidence of injuries in its various sports. The early version of the system used paper forms to gather information from volunteer institutions. The system has evolved to a Web-based program with the goal of following all sports in all institutions. The system allows comparison of injuries among sports, in practice and game situations, in spring versus fall football, on various playing surfaces, and so forth. The surveillance system tracks injury data from year to year, providing the opportunity to search for trends. The system also allows analyses of changes in injury rates following equipment modifications, rule changes, and other factors. The institutions' athletic trainers forward current confidential information to the NCAA's Web site, where it is entered into the system. The organization's Sports Medicine Committee studies the data, uses them to determine the effectiveness of rule or equipment changes, and uses the system to determine areas in need of study.

Analysis: Quantitative and Qualitative Information

To analyze an evaluation, employers should begin with the reasons they undertook the evaluation in order to focus the analysis and organize the data. The data should be organized according to the categories of process, impact, and outcome evaluation.

Numerical data can be entered in a spreadsheet (e.g., Microsoft Excel) to allow the calculation of means, standard deviations, ranges, and percentages. Spreadsheets allow statistical comparisons of such measures as pre- and posttest (or fitness program) cholesterol and aerobic fitness. They also allow the calculation of relationships (correlations) among variables, such as the relationship of aerobic fitness to cholesterol, blood pressure, or body weight, or changes in fitness with changes in the variables. Although it is acceptable to search for significant changes or relationships, it is more appropriate to include comparisons in the

CASE STUDY
Fitness and Performance Among Wildland Firefighters

A study by Sharkey (1981) of 121 wildland firefighters (97 men, 24 women) involved measures of aerobic fitness, muscular fitness, and body composition, followed by a series of timed performances in job-related field tasks (dig fire line, drag hose, shovel, carry load). All firefighters had met the aerobic fitness standard of 45 (ml · kg^{-1} · min^{-1}). Total time to complete the six tasks was found to be significantly related to measures of aerobic fitness, muscular fitness, and body composition, including lean body weight. The scores for each crew member were averaged to determine an average time per crew. Smoke jumpers, with an aerobic fitness average of 56 (ml · kg^{-1} · min^{-1}), averaged 72.5 minutes on the tasks, and a Hotshot crew with a fitness average of 54 averaged 86.9 minutes. Two crews with fitness averages of 55.6 and 51.6 scored 108.3 minutes and 125.8 minutes on the tasks, respectively.

The skilled smoke jumper and Hotshot crews were highly motivated during the trials, whereas the other two crews were inclined to "go through the motions" in this simulated work environment. A less experienced women's crew, with a fitness average of 48.5, scored 171.4 minutes on the six tasks, over two times slower than the smoke jumpers. The women's average lean body weight was 101.5 pounds (46 kg), whereas the men's was 152 pounds (69 kg). The correlation of LBW to performance on the tasks (.61) was statistically significant.

This study illustrates the importance of motivation in human performance, and it suggests the need to consider cultural and psychosocial factors in the evaluation. The crew with the fitness score of 55.6 came from a culture with a different concept of work performance. They approached the field tasks as they would any other day's work, pacing themselves rather than competing with each other or other crews. This approach works well on the job, but it does little to illuminate the relationship of aerobic fitness to work performance.

Subjective rankings of attitudes, job satisfaction, and morale can be viewed as qualitative information. They can be quantified and analyzed on the spreadsheet. Verbal or written responses to interviews and questionnaires can be organized into categories, such as concerns, suggestions, strengths, weaknesses, and recommendations. Employers should seek evidence of patterns in the information, such as more favorable responses from those who scored higher or participated more often.

The analysis and report should begin with quantitative data. Employers should put the information in perspective, as in a scientific study, identifying the hypotheses (what was expected) and then showing what actually happened. They can use qualitative information to flesh out the findings and to put a human perspective on the testing and fitness programs. The number of people involved, how many passed, the reason for failure, and differences among groups should be indicated. To evaluate a fitness program, employers should use quantitative data to show the effect on participants and use qualitative data to improve program delivery.

Remember that all medical and fitness information should remain confidential. The program evaluation should not include names or identifying numbers. In the NCAA injury surveillance system, names are not entered in the database, and access to institutional data is limited so that outsiders cannot infer the injury status of opponents. U.S. federal law (HIPPA) mandates that medical information remain confidential. Failure to maintain that trust will undermine confidence in the program.

Cost-Benefit Analysis

The cost-benefit analysis requires an examination of certain variables in relation to the cost of program implementation. Variables to consider include productivity, absenteeism, attrition, employee morale, use of health care, and psychological services. The first step is to identify the costs of program implementation. The next step is to examine the variables to determine the value of the benefits.

For a job-related test, employers should determine the salary costs of the program, extra medical costs (if any), facility rental, consultants, printed materials, advertising, mailings, pretest training, and other expenses. They may include start-up costs, including the cost of the job task analysis, test development, and validation, but they should keep in mind that the program should not be expected to pay itself off in the first year. For a job-related fitness program, salaries for administrators and employees in the program, additional medical costs, certified fitness instructors, facility costs, and so on, should be determined. Then it is time to search for the benefits.

Individual productivity is hard to measure if the job is not easily quantified. Performance ratings by supervisors may be used, but these are subject to bias, in favor of some and against others. If possible, employers should use the average of several ratings. Team or crew performance can be inferred by subject matter experts in the field. Experienced supervisors can judge quality performance, how it has improved, and those workers responsible for the improvements. However, when performance is measurable, employers should measure it. In the *SEPTA* case individual productivity was measured using arrest rates and commendations. Team productivity was measured according to reductions in certain types of felonies.

Absenteeism is easily quantified, although some cases (such as that of a sick child) can be ignored in the analysis. Cases of attrition should be studied to determine the cause. For example, was the job too difficult, was management too demanding, were coworkers not supportive, or were there physical or psychological problems? The costs of recruitment and training are considerable, making employee turnover an important consideration in the cost of doing business. Employee health and fitness programs can reduce employee turnover by improving morale and camaraderie. Employee morale can be assessed with an interview or questionnaire. Job satisfaction, a sense of accomplishment, and career progress contribute to morale and productivity.

Employees' use of health care and psychological services can be used to assess a program's effectiveness. This confidential analysis considers the number of employee visits, health care costs, and expensive procedures to determine the use of health care. Implementing a physical fitness program may result in a decline in the use of health care and psychological services. A health and injury surveillance system can facilitate an analysis of benefits.

Ultimately, the cost-benefit analysis comes down to a comparison between two columns of numbers.

Employee morale can be assessed with an interview or questionnaire.

Unfortunately, costs are easier to calculate than benefits. A significant decline in group cholesterol, blood pressure, or BMI is good for employees, but the value to the organization (including the cost of replacing seriously ill or deceased employees) is difficult to determine. However, if reducing coronary artery disease risk is one of the goals, the program is having a positive impact.

Reporting Results

The evaluation report attempts to put the findings in perspective and compares results to what was hypothesized and to program goals. It lists positive outcomes and costs and benefits and can provide recommendations to improve the program. It should begin with an executive summary. The report should then describe the program being evaluated, the goals of the evaluation, and the methods and analysis procedures and end with conclusions and recommendations. Charts and graphs are useful for describing findings. A written report can be supplemented with a PowerPoint presentation. It should be distributed to managers and employees, and a time for discussion should be scheduled.

When evaluation data are analyzed by those responsible for the test or program, objectivity may be compromised. If management or employees are concerned about a fair evaluation, outside help can be obtained. At the very least, someone other than the program leader should review the procedures, conclusions, and recommendation.

Appointing an evaluation working group composed of the program leader, an employee, and a manager may be helpful. Involving this group early in the process could ensure acceptance of the final report.

The evaluation working group can plan, carry out, and report the findings. Members of the group should be tactful and aware of the history, purpose, and field operation of the program. Managers and employees can guide the process by focusing on important questions. Management should set the scope, time lines, and completion date. The working group may decide to engage the services of specialists to help focus data collection and analysis. Preliminary findings can be discussed with key members of the organization.

It is important to plan the evaluation process before the fact to ensure that appropriate sources of data have been identified and collected. There is no single way to conduct an evaluation. Those involved should agree on a plan and begin collecting data. They should include subjective interviews and questionnaires along with objective numerical data to determine benefits and to indicate areas in need of improvement. Organizations learn a great deal about their programs by understanding reasons for failures and dropouts. All evaluation materials should be kept on file for future reference. Success comes when the program is open to analysis and constructive criticism. Evaluation provides the feedback needed to identify necessary changes.

Summary

In this chapter you learned how to use program evaluations and when to consider the use of additional measures. You learned about process, impact, and outcome evaluations and when to use each based on the goals of the process. A process evaluation answers questions about numbers, demographic variables, pass/fail rates, participation rates, fitness program attendance, and satisfaction. An impact evaluation examines the health and behavioral changes that occur following participation in the program. Finally, an outcome evaluation examines the changes that take place in the organization and considers the cost of programs in relation to their benefits. Data on productivity, absenteeism, attrition, employee morale, training, workers' compensation costs, and costs of health care are used to determine the cost-effectiveness of the program.

You also learned to use appropriate statistical techniques to analyze quantitative data. Sophisticated statistical techniques are seldom needed in an evaluation. An analysis of an evaluation should list hypotheses, state what was expected, and then show what actually happened. It should indicate how many people were involved, how many passed, reasons for failures, and differences among groups.

An evaluation report should begin with an executive summary and then list hypotheses and summarize findings, beginning with quantitative data. Qualitative information can be used to flesh out the findings and put a human perspective on the program. The costs and benefits of the program should be summarized, and the report should end with conclusions and recommendations.

The evaluation process is useful for determining the costs and benefits of testing and fitness programs. The costs and benefits of employee health and wellness programs and medical standards and examinations should also be evaluated. The evaluation process provides the data that employers need to identify positive outcomes and to fulfill the need for necessary changes.

PART IV

Employee Health, Physiology, and Performance

© Human Kinetics

The reason why worry kills more people than work
is that more people worry than work.

Robert Frost

Many Americans have come to rely on their doctors to take care of their health. Of course, it doesn't work; our doctors can't make us stop smoking, lose weight, eat less fat, fasten our seat belts, or get regular exercise. These simple habits are more related to health and disease than all the influences of medicine. More than half of all diseases and deaths can be attributed to the way we live our lives. We have been led to believe that an annual medical examination will help us stay healthy and reduce the likelihood of illness or premature death. But when researchers compared those who had annual exams with those who did not, they found similar incidences of chronic disease and death in both groups.

Similarly, employees in certain occupations rely on medical examinations, avoiding efforts to improve their health habits such as regular physical activity. Structural firefighters have developed extensive medical standards that are recommended for all employees (NFPA 1582), but they have avoided mandatory fitness programs. Instead, they have a voluntary health and fitness program (NFPA 1583) and alarming incidences of coronary artery disease. Chapter 10 considers the values and limits of medical examinations and employee health and fitness (wellness) programs. We suggest ways to lower medical costs while improving the health of employees. Chapter 11 addresses work physiology and its relationship to performance, and chapter 12 presents the essential element required to preserve career-long performance and health: the job-related physical fitness (or physical maintenance) program.

Chapter 10
Employee Health

© Human Kinetics

A report in the *Archives of Internal Medicine* concludes:

"Avoiding a sedentary lifestyle during adulthood not only prevents cardiovascular disease independently of other risk factors but also substantially expands the total life expectancy and the cardiovascular disease-free life expectancy for men and women. This effect is seen at moderate levels of physical activity and the gains in cardiovascular disease-free life expectancy are twice as large at higher levels of activity" (Franco et al. 2005, p. 2358).

This chapter considers the components of the employee health program, including medical standards, the medical examination, and the health education (wellness) program. We describe the components and discuss their benefits and their costs.

This chapter will help you do the following:

- Develop medical standards and a medical examination plan.
- Outline a comprehensive employee health program that includes health education and health-related fitness.
- Evaluate the cost-effectiveness of the employee health program.

Developing Medical Standards

Establishing medical standards is important for occupations that are arduous or hazardous. The medical standards and examination should consider the risks and the functions associated with employees' duties and responsibilities. Candidates should be medically evaluated and certified by the department's physician or by a board-certified physician with extensive knowledge of the occupation and its demands. Incumbents should be evaluated periodically (every three to five years) or as determined by laws and regulations. The medical examination should be viewed as part of the employee health program.

Medical standards are provided to aid the examining physician, the designated medical review officer, and officials of the department when determining whether medical conditions may hinder a person's ability to safely and efficiently perform the requirements of the occupation without undue risk to self or others. The standards also serve to ensure consistency and uniformity in the medical evaluation of applicants and incumbents for arduous or hazardous occupations. Medical standards should be subject to clinical interpretation by a medical review officer with extensive knowledge of the job requirements and environmental conditions in which employees work. The medical review officer ensures the consistent and uniform application of the standards. This is especially important when a number of examining physicians are involved and candidates are evaluated in several locations.

When faced with the decision of whether to hire someone with a preexisting medical condition, the employer must document the degree of impairment; weigh the seriousness of the applicant's medical condition; identify medication requirements; and consider whether the functional requirements of the job and working conditions would aggravate, accelerate, or permanently worsen any preexisting medical condition.

Minimum Medical Qualifications

The medical standards establish the minimum medical qualifications necessary for safe and efficient job performance. The initial determination of minimum medical standards should be evaluated over time and adjusted as necessary. The standards cover the following health- and performance-related areas:

Vision: Visual acuity, binocular vision, depth perception, peripheral vision, and color vision. The standard covers the use of corrective and contact lenses.

Hearing: The ability to hear verbal communication, both natural and manufactured warning sounds, and radio communication. The standard sets auditory limits and covers the use of hearing aids.

Immune system: Freedom from infectious diseases or immune system or allergic conditions that could compromise health or performance.

Head, nose, mouth, throat, and neck: The ability to move the head, breathe freely, wear personal protective devices, and communicate.

Dermatology: Intact and healthy skin that allows exposure to working conditions.

Cardiovascular system: A healthy vascular system with no phlebitis, thrombosis, or arterial insufficiency and without hypertension (with or without medication); a healthy heart with a low risk of cardiac disorders and coronary artery disease and with no electrocardiogram abnormalities (for those over 40-years of age).

Respiratory system: A healthy respiratory system, lung function, and residual capacity; capable of using a respiratory protective device.

IIIIIIIIIIIIIIIIIII MEDICAL IIIIIIIIIIIIIIIIIIII STANDARDS FOR WILDLAND FIREFIGHTERS

Medical standards were developed for wildland firefighters whose jobs are arduous and hazardous and performed under variable and unpredictable working conditions (Jensen 2005). The standards were developed to ensure that firefighters will be able to perform the full range of requirements of their duties under the conditions under which those duties must be performed; to ensure that existing or preexisting medical conditions of firefighters and applicants will not be aggravated, accelerated, exacerbated, or permanently worsened by their work tasks; and to demonstrate the fire community's strong commitment to public and employee health and safety.

Endocrine and metabolic system: Capable of function under normal and emergency working conditions.

Blood: A healthy blood and blood-producing system capable of performing in the occupation's working environment.

Musculoskeletal system: Adequate strength, muscular endurance, flexibility, and range of motion and lacking preexisting joint or back problems.

Nervous system: A healthy sense of balance and sensation of the environment and oneself and a low risk of sudden or subtle incapacitation.

Gastrointestinal system: The ability to consume adequate calories, nutrition, and fluids and a low risk of sudden or subtle incapacitation.

Genitourinary system: A healthy system with a low risk of sudden or subtle incapacitation.

Mental health: A healthy sense of judgment, healthy mental functioning and social and behavioral skills.

Pregnancy: Although not a disability, pregnancy should be identified if work is performed in environments that could affect fetal growth and development or maternal health.

Medical standards should also address the acceptability of prosthetics, transplants, and implants, such as pacemakers and defibrillators. Each standard should have acceptable levels and conditions that may disqualify participation. For example, a knee or hip replacement or a coronary bypass operation may disqualify participation, whereas pregnancy may lead to a temporary change in duty status. Once the standards have been determined, it is time to determine objective medical examination criteria.

Medical Examination

The details of the medical examination are beyond the scope of this book. However, we do recommend that objective measures be used to determine critical standards. Blood pressure and blood lipid measurements are inexpensive and accurate measures of cardiovascular risk. Given that medical and workers' compensation records indicate the high cost of preexisting knee and back conditions, it makes sense to use proven, objective criteria of function. Knees should be observed in action, climbing and descending steps, and in one-leg knee bends, with and without loads. Backs can be evaluated by having the person bend, twist, and move an occupationally relevant load. Job-related

CASE STUDY

Medical Standards for Arduous Work

Five U.S. federal agencies decided to consolidate their medical standards for arduous work. A committee was formed, consisting of a personnel specialist, a lawyer, several field personnel, and a physician (Jensen 2005). Members of the committee underwent training for the job, made field visits to observe workers, and reviewed medical standards adopted by similar organizations. They then began the task of developing the medical standards. Various functional and systematic categories were created, and each category was assigned a standard. Categories included vision, hearing, musculoskeletal, cardiovascular, respiratory, digestive, nervous system, and mental health. Once the committee members agreed on the extensive standards, they began the task of creating forms to guide the physicians conducting the examination. Finally, they developed guidelines that would govern the process, including scheduling, waivers, accommodations, and medical review.

The standards received tentative approval and were field tested on 2,000 employees. When management discovered the hefty price of the comprehensive examination, it asked the committee to identify ways to reduce the cost without sacrificing the health of employees. On reconsideration, the committee decided to eliminate several high-cost items: an expensive blood test for pesticide exposure, the chest X ray, and the resting electrocardiogram for employees less than 40 years of age. The revised and less costly examination was approved, and the standards are now in effect.

aerobic and muscular fitness can be evaluated in the job-related work capacity test.

Medical examinations for applicants may be more comprehensive than those for incumbents to avoid long-term medical and rehabilitation costs. Exams for incumbents could be scaled to meet the major risks. Employers should avoid spending money on expensive tests, such as an audiometer, when a less costly evaluation (the whisper test) may do. Workers exposed to loud equipment (e.g., chain saws) will still be tested and required to meet Occupational Safety and Health Administration (OSHA) hearing standards in the United States.

When 2,000 wildland firefighters were tested for vision, almost 200, or 10 percent of the candidates, did not pass the 20/40 standard. All of the candidates who failed sought a waiver through a formal administrative appeal. Because all the workers had been employed previously, all wore glasses, and all of them met the standard with their glasses, they were passed with the proviso that they carry an extra pair of prescription glasses. It might be more cost-effective to tell those with glasses to bring an extra pair, and to avoid the costly vision test, and the time-consuming administrative appeal.

Employers should examine incumbents every three to five years until 40 or 45 years of age; then every two or three years thereafter. The periodic examination should be altered to fit the age group. A resting ECG should be added at 40 years of age and a PSA (prostate-specific antigen) test at age 50 (for men). In the years between medical examinations, incumbents should complete an updated health history, indicating any changes that have taken place. Blood pressure and information from the annual health screening program can be used as part of the annual evaluation (see page 102).

Because medical examinations are very expensive, conducting a cost-benefit analysis when selecting tests is recommended. There is little reason to include a chest X ray (and evaluation by a radiologist) in the medical examination. A nodule observed on an X ray requires a biopsy, and few nodules are malignant. In addition, the X ray itself could increase people's overall cancer risk. Tests should fit the demographics and risks of the population. According to the American College of Cardiology, the resting electrocardiogram is an appropriate test for people 40 years of age or older (Gibbons et al. 2002). Tests on young recruits are likely to indicate more false positive than true positive results, causing concerns for healthy workers. And, except for those with symptoms, risk factors, or a family history, the same is true for the treadmill "stress" test. In other words, it is not cost-effective.

Evaluation

It's a good idea to use the evaluation process outlined in chapter 9 to determine the costs and benefits of the medical examination and to judge the contributions of standards and tests. The threat of a lawsuit forces many organizations to conduct unnecessary or expensive tests, thereby practicing so-called defensive medicine. For example, in the world of sport, athletes occasionally die abruptly from cardiac abnormalities such as hypertrophic cardiomyopathy or Marfan's syndrome. Although the incidence of these fatalities is quite low (fewer than 1 in 200,000 athletes), and the cost of testing is high, some argue that all athletes should be evaluated before they train and compete. A careful history and examination help identify those at risk. Routine use of the 12-lead electrocardiogram, echocardiography, or graded exercise test is not recommended in large populations of athletes (Maron et al. 1996).

In the workplace some argue for the use of a treadmill test for recruits and incumbents. This test can indicate previously undiagnosed heart disease, but heart disease isn't a big problem in young recruits and incumbents. The treadmill test is better suited for workers who have symptoms, several risk factors (elevated blood pressure, cholesterol, overweight, cigarette smoking, over 45 years of age), or a family history. When

The resting electrocardiogram is an appropriate test for people 40 years of age or older.

© ASSOCIATED PRESS/SARA D. DAVIS

The treadmill test is best suited for workers who have symptoms or several health risk factors.

the test is performed on young, active people, false positive results increase from 10 percent to over 30 percent (American College of Sports Medicine 2005). A positive result from a treadmill test means that the person may have coronary artery disease. A person who receives a false positive result will think he has heart disease until he takes one or more tests to verify the findings. In some cases an invasive coronary angiogram is needed to determine whether the person needs drug therapy, coronary angioplasty, or a bypass operation. The point here is that employers should use tests that are appropriate to the age and activity level of the population.

Medical Surveillance

Information from the entry-recruit and incumbent medical examinations will aid in medical surveillance. Surveillance means keeping a close watch on measures that could be affected by the demands and conditions of employment in the physically demanding occupation. In fields in which coronary artery disease is a problem, employers should keep track of heart disease risk factors. Many think that firefighters should receive periodic pulmonary

CASE STUDY
Philadelphia Fire Department

Researchers conducted a retrospective cohort mortality study of 7,789 firefighters employed by the Philadelphia Fire Department between 1925 and 1986 (Baris et al. 2000). The average length of employment was 18 years, and all employees had spent some part of their careers fighting fires. The project analyzed mortality patterns, focusing particularly on cancers that could be caused by by-products of combustion. Firefighter death rates were compared with those of the general population of white males of the same age (the majority of the firefighters were white men). Standardized mortality ratios (SMRs) and confidence intervals were computed. An SMR of 1.0 is identical to the comparison population.

There were 2,220 deaths of known origin among the firefighters. The cohort had similar mortality from all causes of death combined (SMR = 0.96) and all cancers (SMR = 1.10). Statistically significant excess risks were observed for colon cancer (SMR = 1.51) and ischemic heart disease (SMR = 1.09). Elevated, although not statistically significant, risks were

found for cancers of the pharynx and bladder, non-Hodgkin's lymphoma, and multiple myeloma. The number of observed cases of most other cancers, including lung cancer, and nonmalignant causes of death were similar to or slightly lower than that of the comparison population. There were statistically significant deficits of death from nervous system diseases, cerebrovascular diseases, respiratory diseases (SMR = 0.67), genitourinary diseases, all accidents, and suicide. In general, SMRs were higher among firefighters 65 years of age and older.

This study and others document the risk of ischemic heart disease in firefighters, a surprising finding in a relatively young and active cohort. The firefighter fatality retrospective study conducted by the U.S. Fire Administration documented the nature of fatalities in nonwildland and wildland firefighters. Although in both groups 1 percent of deaths resulted from stroke and cerebrovascular events, the nonwildland firefighters had almost seven times the number of cardiac arrest/heart attack deaths of the

(continued)

(continued)

wildland firefighters (47 percent versus 7 percent) (United States Fire Administration 2002). A National Fire Protection Association study of U.S. firefighter fatalities due to sudden cardiac death reported 440 victims over the 10-year period 1995 to 2004 (Fahy 2005). Of that total, 307 (69.8 percent) were volunteer firefighters, 117 (26.6 percent) were career firefighters, and the remaining 16 (3.6 percent) were

wildland or military firefighters. When the National Fire Service Research Agenda Symposium (2005) considered risk factors relating to cardiovascular disease in the fire service, it recognized the effect of diet and exercise on heart disease risk factors and suggested that a commitment to health and wellness programs could lead to significant reductions in cardiovascular deaths and disabilities.

function tests, but the Philadelphia Fire Department study (see the case study) did not indicate an elevated incidence of respiratory disease or lung cancer in firefighters. OSHA requires that those working with loud equipment have regular hearing tests and wear hearing protection. Employers should track illness or injury trends to gauge the impact on employees of changes in equipment or tactics.

Employee Health Programs

The real key to worker safety, health, and productivity is an effective employee health and wellness program. Employee health programs have expanded as the costs of medical care and health insurance have risen dramatically. The employers' percentage of the cost of U.S. health care has exploded to over 30 percent, contributing to labor costs, layoffs, and even bankruptcy. The employee health program is a combination of activities designed to improve employee health via health education, risk factor reduction, and health-related fitness. It includes educational, organizational, and environmental activities designed to motivate healthy behaviors among employees and their families (Chenoweth 1998).

The employee health program consists of primary and secondary prevention and, in some organizations, tertiary care. Primary prevention begins with a health risk analysis, health screenings, and immunizations, and it may also include ergonomic analyses. Secondary prevention includes health education (nutrition, stress reduction), health fairs, health-related fitness, and medical examinations. Tertiary care covers rehabilitation and case management. Some programs offer child care, on-site medical care, a fitness facility, even access to a dry cleaner. Although the program sounds expensive, studies show that a comprehensive program saves

$3.50 to $6.00 for every dollar spent (Pronk 2003). Organizations offer employee health programs to reduce absenteeism and workers' compensation costs, lower health costs and insurance rates, increase productivity, retain employees, and compete for new ones. See an example of an employee health program on page 106.

Employee Health Program

Job-Related Components	Health-Related Components
Medical standards	Health risk analysis
Medical examination	Health screening
Medical surveillance	Immunizations
Ergonomic analysis	Health education*
Rehabilitation	Health-related fitness**
Job-related fitness	

*Nutrition, stress management, weight control, back health, smoking cessation.

**Aerobic, muscular, flexibility, body composition.

Health Risk Analysis and Health Screenings

The health risk analysis (HRA) is a computer-scored appraisal that identifies health risks and ways to reduce them. Most forms compare the person's chronological age with her risk age. An obese smoker could have a risk age well above her chronological age, whereas an active, lean, nonsmoker could have a risk age 10 or more years below her actual age. The HRA is a very low-cost motivational tool (Sharkey and Gaskill 2007).

Once- or twice-a-year health screenings keep track of risk factors. Screenings usually include body weight and blood pressure measures; blood tests (for lipids, glucose, liver function measures, and others); prostate-specific antigen (PSA) tests to screen for prostate cancer; dermatological exams; and so forth. Flu shots and other immunizations should be offered during the fall session. These low-cost tests can be conducted in the workplace

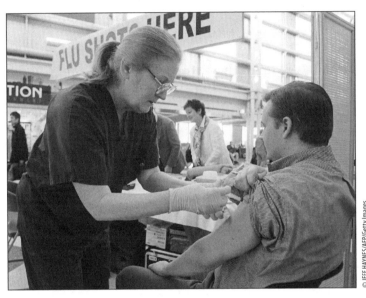

© JEF HAYNES/AFP/Getty Images

A flu immunization is an important component of primary prevention in an employee health program.

at a fraction of the cost of tests at a medical facility, and they reduce the need for visits to the doctor's office.

If an employee's age, sex, race, family history, health habits, or exposures put him at risk for a disease, by all means an employer should use an early detection test, as long as the proposed test meets the following criteria recommended by the World Health Organization (see www.who.dk):

- The disease has a significant effect on the quality of life.
- Acceptable methods of treatment are available.
- Treatment during the asymptomatic (no symptom) period significantly reduces disability or death.
- Early treatment yields a superior result.
- Detection tests are available at a reasonable cost.
- The incidence of the disease in the population is sufficient to justify the cost of screening.

|||||||||||||||||||||||||||||||||||||||||| HAZARDS OF SMOKING ||||||||||||||||||||||||||||||||||||||||||||

No fire department should permit smoking. The U.S. surgeon general's report, *The Health Consequences of Smoking: Cancer and Chronic Lung Disease in the Workplace* (1985), notes that asbestos exposure increases the risk of lung cancer 5 times, smoking increases the risk 10 times, and smoking and asbestos exposure increase the risk 87 times (www.surgeongeneral.gov/library/reports.htm). Everyone knows about the health hazards of smoking, including lung cancer, coronary artery disease, stroke, and chronic obstructive lung diseases such as emphysema and chronic bronchitis. But those hazards are only part of the problem. Recent research has added depression and impotence to the list of problems caused by smoking.

Studies show a relationship between tobacco use and illness, injury, and long-term disability. Significant correlations were observed between smoking and injury, total lost workday injuries, and illnesses for manual material handlers in the United States (McSweeney et al. 1999). Musculoskeletal disorders

were associated with smoking in U.S. automotive workers, and the problems increased with exposure (White 1996). Smoking predicted long-term disability in less active workers, and musculoskeletal pain was more likely among construction workers who smoked a pack or more a day. Smoking seems to affect blood vessels and blood flow, leading to overuse injuries, low back pain, and long-term disability. It also leads to reduced perfusion and malnutrition of tissues in or around the spine and causes these tissues to respond poorly to mechanical stress (Eriksen, Natvig, and Bruusgaard 1999). Army studies have documented higher musculoskeletal injury rates among smokers (Amoroso et al. 1996). Smokers are believed to be risk takers, heavier users of alcohol, and more frequently diagnosed with depression. They have impaired healing of wounds and susceptibility to fractures. Smoking also has been linked to overuse injuries, motor vehicle accidents, industrial accidents, low back and shoulder pain, burns, and suicide.

Few early detection tests meet these criteria; tests of blood pressure, breast examinations, and Pap smears do meet the criteria, whereas exercise stress tests, diabetes screenings, and X rays for lung cancer do not. Surprisingly, the routine (annual) medical examination also fails to meet the criteria. Employers should not use tests just because they are available.

Health Education

A needs assessment of employee interests will determine the components of health education. Most education programs include stress management, nutrition, weight control, back health, and smoking cessation. Some offer medical self-care, parenting, and prenatal classes.

Stress is in the eye of the beholder; law enforcement is exciting to some and stressful to others. Jobs in and of themselves are not stressful; however, our reactions or responses to jobs may be stressful. Stress is defined as a rise in adrenocortical hormones, such as cortisol. Because prolonged exposure to cortisol leads to diminished immune function and other health effects, employees must learn to manage stress. For a comprehensive look at stress management, see the *Stress Management* video from Chuck Corbin's five-video *Fitness for Life Video Series: Wellness* (Corbin 2005).

Nutrition involves making food choices that ensure the energy, vitamins, and minerals required for health and performance. The best nutritional strategy for promoting optimal health and reducing the risk of contracting chronic diseases is to obtain adequate nutrients from a wide variety of foods. Whole foods are the best sources of vitamins and minerals. Supplements are not a substitute for a balanced and nutritious diet. For more information, go to www.mypyramid.gov/ and to www.health.gov/dietaryguidelines/dga2005/. Another good source is *Healthy Eating Every Day* by Carpenter and Finley (2005).

Weight control is the balance of energy intake and energy expenditure, of matching caloric intake to the number of calories burned. To lose weight, people should reduce caloric intake, increase caloric expenditure, and—if needed—employ behavior modification techniques. For more information on weight control, see *Fitness and Health* (Sharkey and Gaskill 2007).

Back health is maintained with regular attention to good posture; weight control; and flexibility, abdominal, and back exercises. Those with back problems need to return to these exercises as soon as possible. Core and muscular fitness training help protect the back from injury. For more information, see *Fitness and Health* (Sharkey and Gaskill 2007).

Smoking and substance abuse are difficult problems that are best dealt with in targeted programs that involve group or individual counseling.

Health-Related Fitness

Health-related fitness differs from job-related fitness, but the two are complementary. Job-related fitness requires more attention to muscular fitness and job-related aerobic training. Regardless of their occupation, all employees should take part in health-related fitness programs. Regular exercise reduces the risk of heart disease, stroke, diabetes, obesity, and some cancers. Active employees are less likely to smoke. Those who engage in health-related fitness training (currently less than 20 percent of the adult workforce) pay several thousand dollars less annually in health care costs. Active people also outperform their inactive coworkers. The goal of a health-related fitness program is to improve the health and the quality of life of all employees.

A word of warning is appropriate at this point. When we refer to a physical fitness program, we mean carefully developed and applied prescriptions for aerobic and muscular fitness. We do not mean recreational volleyball, basketball, or soccer. These activities can cause injuries and increase health care costs.

Health-related fitness consists of the following:

Aerobic fitness: Regular participation in meaningful aerobic activities with the goal of meeting the American College of Sports Medicine and the U.S. Centers for Disease Control standard, which is 30 minutes of moderate activity per day most days of the week (Pate et al. 1995; Haskell et al. 2007), or the Institute of Medicine goal of 60 minutes per day (Institute of Medicine 2002).

Muscular fitness: Sufficient strength, muscular endurance, and flexibility to carry out daily tasks with vigor and alertness and with the capacity to meet unforeseen emergencies. All employees should engage in core training to maintain back health.

U.S. Navy photo by Mass Communication Specialist 2nd Class Joseph M. Buliavac
http://www.navy.mil/view_single.asp?id=42792

Sailors on board USS *Ronald Reagan* run on newly installed treadmills in the mezzanine gym. All employees should take part in health-related fitness programs because regular exercise reduces the risk of heart disease, stroke, diabetes, obesity, and some cancers.

Body composition: A reasonable body weight (BMI 19 to 25), a healthy waist-to-hip ratio, and a healthy girth at the waist, are indicators of a healthy body composition.

The health-related fitness program can be pursued at an on-site fitness center, at a commercial health club, or on the employee's own time. Health fairs and periodic tests can motivate participation. Although some organizations use monetary incentives to encourage participation, long-term involvement depends on intrinsic motivation (doing it because you want to), essential for health and quality of life. For example, Round and Green (1998) investigated the impact of monetary reward on firefighter participation in a fitness program. Twenty-six male firefighters started the study. Participants averaged 33.3 years of age, 195.9 pounds (88.9 kg), and 19.5 percent body fat. Firefighters received $100 a month for averaging above the 80th percentile in several fitness tests. That is $1,200 in one year! Most folks would love to have that opportunity, and yet the results indicated that the money wasn't sufficient motivation for participation. In fact, only 18 of the eligible firefighters completed all testing sessions. Previous studies of monetary incentives revealed more success in programs that had more structure than the Round and Green program. The authors of this study felt that the participants may have had a concern for job security if higher standards were required. In

a discussion of this study with fire and union personnel, we asked why the firefighters did not respond to the monetary incentive. To them the answer was obvious: They were too busy with second jobs. Their 24-hour work schedules allowed several days off every week, and they filled the spare time, not with volunteer work, recreational activities, or fitness training, but with additional employment.

If monetary or other rewards are to be employed, they should be based on excellence in performance, not on sustaining a minimum level of fitness commensurate with the essential functions of the job. We do not recommend rewards for showing up or taking a test. We are not enthusiastic about providing paid time for fitness training. But then, we are old-fashioned; we engage in fitness training for its own sake, regardless of the requirements of the job.

Evidence from around the country indicates that the majority of firefighters and law enforcement personnel do not participate in health-related fitness programs, or even job-related fitness programs. Indeed, their unions have been resistant to the implementation of mandatory fitness requirements or annual evaluations of work capacity. This is most surprising in firefighting, a physically demanding job with ample on-the-job time for training. If money won't work and employees refuse to participate, what is the answer?

One way to address the issue of fitness requirements for public service personnel is to discuss it at city council or town meetings. These civil servants are well paid, and their jobs include health insurance and retirement benefits. Taxpayers invest a lot in their fire and police departments. They purchase expensive vehicles and top-of-the-line personal protective equipment. They invest in medical examinations, and yet employees have a surprisingly high incidence of heart disease. Citizens need to know how little many employees are willing to invest in their jobs, in their ability to serve and protect. A public discussion could help focus attention on factors related to safety, health, and performance.

Costs and Benefits of Employee Health Programs

Evaluating employee health programs is important, especially the parts that cost the most. Employers should use health care and injury data to evaluate the costs of the medical examination. Money should not be spent on tests that yield little useful information, tests that produce lots of false positive results, or tests whose costs exceed their benefits. Employee health care costs and illness (morbidity) and fatalities (mortality) data can reveal major problems. Employers should select exam elements that identify risk factors. If possible, they should try to avoid the practice of defensive medicine.

Figure 10.1 illustrates the relationships between the benefits and risks of physical activity. The figure indicates that benefits increase rapidly at first but then begin to plateau, with little additional reward at higher levels of activity. Risks, on the other hand, rise slowly at first, then more rapidly at higher levels of activity. It seems prudent for workers to

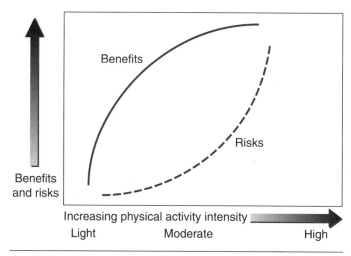

Figure 10.1 Relationships of benefits and risks with levels of physical activity.

Reprinted, by permission, from B.J. Sharkey and S.E. Gaskill, 2006, *Fitness and health*, 6th ed. (Champaign, IL), 59; adapted, by permission, from K. Powel and R. Paffenbarger, 1985, "Workshop on epidemiologic and public health aspects of physical activity and exercise: A summary," *Public Health Reports* 100 (2):118-126.

CASE STUDY

The Irving Fire Department Wellness Program

The Irving Fire Department Wellness Program has been in effect for 11 years. In 1994 a firefighter task test (the same test given for the entrance exam) was given to all incumbent members of the department to measure their ability to safely execute the recognized job tasks. In 2000 the medical/physical component was added, bringing the program into compliance with National Fire Protection Association (NFPA) and OSHA regulations.

The two components are given on alternating years, the task test one year and the medical/physical test the following year. These two components provide information regarding firefighters' ability to perform their duties as emergency responders. The third component of the wellness program is the time allotted for exercise and the hardware available to achieve the results. The department has 10 firefighters trained as personal trainers, a full complement of weights, and limited pieces of aerobic equipment at the stations. These three components have worked together to provide the following results: In 1994 approximately 220 members were tested and 20 failed the test (9 percent). Six members were over 50 years old. A third of the members passing the test required 10 minutes or more of rehabilitation time after completion of the test. In 2004, 250 members were tested and 3 failed (1 percent). At that time 50 members were over 50 years old, and no members required any significant rehabilitation.

The City of Irving stated that national statistics for industrial wellness programs indicate that a department can expect a 50 percent drop in time lost per injury following the implementation of a wellness program. Data on the Irving Fire Department indicated that by 2004 the program had produced the predicted results, a 50 percent decrease in time off per injury, from an average of 72 days off per injury in 1996 to an average of 35 days lost in 2004. In 2004 the department responded to a record high of over 18,000 alarms and still maintained a lower average number of days off per injury than it did prior to implementing the wellness program. Several factors have made the program indispensable. Aging buildings, increased traffic, older firefighters, and limited staffing have created circumstances in which maintaining funding for monitoring and maintaining the health and physical capacities of firefighters has become an absolute requirement, not a luxury.

Contributed by John Key, Irving Fire Department.

maximize benefits and minimize risks by engaging in a level of activity associated with enhanced health and performance.

Summary

This chapter outlined the elements of a medical standards/medical examination plan. When developing the plan, employers should use the services of a physician certified in occupational medicine, one who is very knowledgeable about the demands of the job, protective equipment, and the working environment. The standards and examination should serve the needs of recruits and employees, but need not use every test available. Although we do not recommend expensive tests for young, low-risk recruits, we do recommend simple objective tests for important elements, such as the back or knees. The plan should balance costs with benefits.

Medical standards and the examination should be part of the employee health program. A comprehensive employee health or wellness program includes health risk analysis, health screening and immunizations, health education, and a health-related fitness program. Employers that can afford only part of the program should begin with health screening and health-related fitness. The complete program has the potential to lower health risks, health care costs, absenteeism, lost-time injuries, and workers' compensation, and to enhance productivity and morale. Successful programs save $3.50 to $6.00 for every dollar spent.

Health-related fitness differs from job-related fitness, but the two are complementary. Time spent in one contributes to the other. Health-related fitness is important for all employees and involves aerobic and muscular fitness and attention to body composition. Most voluntary programs capture the interest of about 20 percent of the workforce, the group that needs it the least. Employers should search for ways to entice the other 80 percent into the program. The job-related fitness program helps employees maintain physical capabilities required on the job. Participation should be mandatory for those involved in physically demanding occupations.

Finally, evaluating all elements of the employee health program is essential, especially the parts that cost the most. Health care and injury data should be used to evaluate the medical examination. Employers should not spend money on tests that yield little useful information, tests that produce lots of false positive results, or tests whose costs exceed their benefits. They should study employee health care costs to determine major problems, then select exam elements that identify risk factors. If possible, employers should try to avoid the practice of defensive medicine.

The value of the health-related fitness program can be determined based on changes in employees' fitness levels and body weight, reduced absenteeism and health care costs, and increased longevity. Employers should consider subjective measures of employee job satisfaction, retention, and morale. Finally, they should calculate the costs versus the benefits of such programs.

Chapter 11

Physiology of Work

© 1998 EyeWire, Inc.

Physically demanding occupations depend on muscular contractions that allow workers to lift and carry loads; climb stairs, ladders, and hills; and use tools. How do these muscles function? What makes them move, what provides the energy for contractions, and what leads to fatigue? The answers to these and other questions are included in this chapter, an introduction to the physiology of work.

This overview of work physiology explains how muscles contract, the sources of energy for contractions, metabolic pathways, and the role of oxygen. The chapter also identifies the respiratory, cardiovascular, and endocrine systems that supply and support the working muscles. Finally, this chapter addresses how training improves the function of muscles and the body's supply and support systems.

This chapter will help you do the following:

- Understand the structure and function of muscle fibers.
- Determine the energy sources for contractions.
- Outline metabolic pathways and the role of oxygen.
- Identify the roles of the respiratory, cardiovascular, and endocrine systems.
- Understand how training improves the function of muscles and support systems.

Muscle Fibers

Humans have three types of muscle fibers: slow-twitch fibers that are efficient in the use of oxygen (slow oxidative, or SO); fast-twitch fibers that can contract with or without adequate oxygen (fast oxidative glycolytic, or FOG); and fast-twitch fibers that use muscle glycogen for short, intense contractions without reliance on a supply of oxygen (fast glycolytic, or FG). Most people average around 50 percent slow-twitch and 50 percent fast-twitch fibers, with 35 percent fast oxidative glycolytic and 15 percent fast glycolytic. Endurance athletes have a higher percentage of slow fibers, and power athletes usually have more fast fibers.

Interestingly, each muscle fiber in a single motor unit is of the same fiber type, slow or fast. In fact, the motor neuron dictates the characteristics of the muscle fiber. If the fiber is consistently recruited for slow work, it takes on the characteristics of a slow-twitch muscle fiber. If it is recruited for fast contractions, it develops the characteristics of fast-twitch fibers (see table 11.1).

Slow-twitch fibers contract and relax slowly, but they are very resistant to fatigue. They have energy sources and metabolic pathways needed for endurance. Fast-twitch muscle fibers contract twice as fast as the slow ones and produce more force, but they fatigue quickly. The fast oxidative glycolytic (FOG) fibers have a bit less endurance than slow fibers but far more than fast glycolytic (FG) fibers, which are usually reserved for short, intense bursts of effort. Endurance training enhances the oxidative capacity of fast-twitch fibers, leading to greater endurance. Strength training improves the strength of both fiber types.

Muscle Contractions

The brain tells the muscles when and how to contract. When the cerebral cortex makes the decision to move, the action is initiated by the motor cortex. Nervous impulses from the motor area are routed via neurons that descend the spinal cord and synapse with motor neurons that leave the cord and pass the message to muscle fibers (see figure 11.1) As they descend, many neurons cross

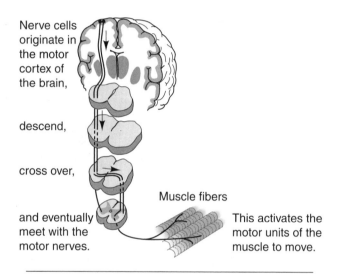

Nerve cells originate in the motor cortex of the brain, descend, cross over, and eventually meet with the motor nerves.

Muscle fibers

This activates the motor units of the muscle to move.

Figure 11.1   The motor cortex and the control of muscles.

Adapted, by permission, from B.J. Sharkey and S.E. Gaskill, 2006, *Sport physiology for coaches* (Champaign, IL: Human Kinetics), 43.

Table 11.1   Characteristics of Muscle Fibers

Characteristics	Slow oxidative (SO)	Fast oxidative glycolytic (FOG)	Fast glycolytic (FG)
Average fiber %	50%	35%	15%
Speed of contraction	Slow	Fast	Fast
Contraction force	Low	High	High
Size	Smaller	Large	Large
Fatigability	High	Medium	Low
Aerobic capacity	High	Medium	Low
Capillary density	High	High	Low
Anaerobic capacity	Low	Medium	High

Adapted, by permission, from B.J. Sharkey and S.E. Gaskill, 2006, *Sport physiology for coaches* (Champaign, IL: Human Kinetics), 46.

to the other side of the cord, which explains why an injury or stroke on one side of the head affects movements on the other side. The motor neuron and the muscle fibers it controls are called a motor unit. Each motor unit includes from a few to many hundreds of muscle fibers.

Each muscle contains thousands of spaghetti-like muscle fibers that range from 1 to 45 millimeters in length. The fibers contain the contractile proteins actin and myosin. Muscle fibers shorten and produce movement when the muscle is stimulated by its neuron. The actin and myosin filaments creep along each other via the tiny cross-bridges that reach out from the thicker myosin and attach to the actin and pull like oars. The barely perceptible movement produced in one segment of the muscle is added to the shortening produced along the length of the fiber, resulting in visible motion (see figure 11.2). Because muscles attach to bony lever systems, their movement is multiplied to produce useful work.

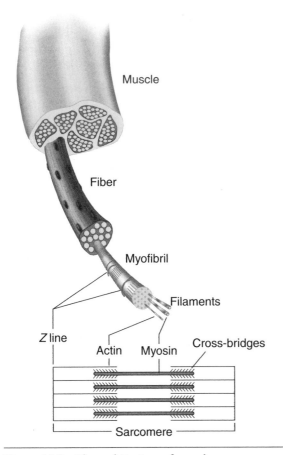

Figure 11.2 The architecture of muscle.

Adapted from B.J. Sharkey, 1990, *Physiology of fitness*, 3rd ed. (Champaign, IL: Human Kinetics), 273.

Force

Because each muscle fiber contracts completely or not at all (all or none), force is dependent on the number of motor units and fibers involved in the contraction. The brain recruits slow-twitch fibers for low force output and larger, fast-twitch fibers for high force output. More frequent nerve impulses can lead to increases in force. Force output declines with fatigue, so more motor units are required to produce the same force.

Fatigue

A fatigued muscle still attempts to contract completely or not at all, but its force declines. Power output declines as intramuscular stores of energy are used (glycogen and fat), acid metabolites increase (lactic acid), energy production drops, and substances involved in contractions (calcium) are depleted. These factors are not unrelated. Increased lactic acid brought on by intense effort alters the cell's acid–base balance and reduces enzymes' rate of energy production, thereby increasing the rate of energy depletion. This is called peripheral fatigue. Proper training can increase energy stores, improve energy production, and reduce the appearance of lactic acid.

Another type of fatigue, central fatigue, occurs when muscle glycogen, the storage form of glucose, becomes depleted after prolonged exertion. The muscle turns to glycogen stored in the liver to supply blood glucose for energy. Eventually, as liver glycogen is depleted, blood glucose levels begin

Central fatigue can be delayed by consuming carbohydrate supplements in solid and liquid form.

to fall. Because the brain and nervous system rely on glucose for energy, the drop in blood glucose (hypoglycemia) is accompanied by impaired nervous system and brain function, confusion, poor coordination, and extreme fatigue. Fortunately, central fatigue can be delayed by consuming carbohydrate supplements in solid and liquid form. Carbohydrate supplements have improved work output of wildland firefighters by 20 percent or more during the latter hours of a 14-hour work shift (Ruby et al. 2004).

Energy Sources

Energy, the ability to do work, comes from the sun, is converted into chemical compounds by plants and animals, and eventually finds its way into the body in the form of carbohydrate, fat, and protein molecules. The chemical breakdown of these molecules, oxidation, releases the stored energy, making it available to power human muscles.

Carbohydrate

Throughout the world carbohydrate provides the major source of energy. It is available in simple and complex forms. Simple sugars such as glucose, fructose, and sucrose (refined sugar, composed of molecules of glucose and fructose) contain energy but few nutrients (i.e., vitamins and minerals). Complex carbohydrates found in beans, rice, whole-grain products (bread, pasta), potatoes, and corn come with important nutrients and fiber. Unfortunately, the average American gets half of his or her dietary carbohydrate from concentrated or refined simple sugars, packed with so-called empty calories (empty because they lack nutrients). Fresh fruits contain simple sugars, but they also provide important nutrients.

Digestion of complex starch molecules begins in the mouth, where an enzyme reduces complex carbohydrates to simple sugars. The process is temporarily halted in the stomach, but then continues in the small intestine, where starches are further digested. Final breakdown to glucose is completed by enzymes secreted by the wall of the intestine. At that point the glucose is absorbed into the bloodstream. The absorption is rather complete; most of the carbohydrate we eat gets into the blood. The absorbed molecules travel to the liver and muscles to restore glycogen stores. Remaining carbohydrate is burned or stored as fat. We recommend that 60 percent of calories come from carbohydrate, especially from whole-grain products. Simple sugars can be used to supplement carbohydrate during long hours of work.

Fat

Fat is the most efficient way to store energy, with 9.3 calories per gram versus the 4.1 and 4.3 calories for carbohydrate and protein, respectively. Dietary fat is broken down and absorbed in the small intestine. It then travels via the lymphatic system, which consists of tiny vessels and nodes that transport and filter cellular drainage. The fat clumps (chylomicrons) are eventually dumped into the bloodstream for transport to cells for energy or to adipose tissue for storage. However, dietary fat intake isn't the only way to acquire this source of energy. Excess carbohydrate or protein can be converted to fat and stored in adipose tissue. There are many ways to acquire fat, but only one good way to remove it: physical activity. We recommend that people limit fat intake to 25 percent of total calories and limit intake of saturated and trans fatty acids (see p. 113).

© Photodisc

Complex carbohydrates are an ideal source of nutrient- and fiber-rich energy and are found in beans, rice, whole-grain products, potatoes, and corn.

FAT INTAKE

If you eat 2,000 calories a day and get 25 percent of your calories from fat, you'll get 500 fat calories. Divide 500 by 9.3 calories per gram of fat, and you get 54 grams, the amount of fat that you can eat each day (far less than the 86 grams you'd eat if fat constituted 40 percent of your caloric intake). Labels on food packages indicate the grams of fat in each serving.

Protein

When we ingest animal or plant protein, the large molecules are cleaved into amino acids and absorbed. The amino acids are building blocks used to construct cell walls, muscle tissue, hormones, enzymes, and a variety of other molecules. Training builds proteins: Aerobic training builds aerobic enzyme protein for energy production, and strength training builds contractile proteins (actin and myosin) for strength. So it should be no surprise to learn the importance of protein to the active life. We recommend that people get 15 percent of their daily caloric intake in the form of protein. Although sedentary people need about 0.8 grams of protein per kilogram of body weight, those in physically demanding occupations need 1.2 to 1.6 grams per kilogram (1.4×80 kilograms [176 lb] = 112 grams of protein per day).

More important than quantity, however, is the quality of protein. Quality protein is high in essential amino acids, those that cannot be synthesized in the body. These essential amino acids are macronutrients, major food sources that must be available for optimal function. When essential amino acids are missing, the body is unable to construct proteins that require them. Although animal protein is a better source of essential amino acids (as well as iron and vitamin B_{12}), proper combinations of plant protein can meet nutritional needs. Protein isn't a major source of energy at rest or during exercise—it seldom amounts to more than 5 to 10 percent of energy needs—but when a person trains hard while dieting to lose weight, the body senses starvation and begins to use tissue protein for energy. To avoid the loss of muscle tissue and to achieve the benefits of training, people should ensure adequate protein and energy intake.

Energy for Contractions

Muscles cannot use carbohydrate and fat directly. They are processed enzymatically to produce high-energy compounds that fuel contractions: adenosine triphosphate (ATP) and phosphocreatine (PCr). When the motor nerve tells the muscle to contract, ATP is split to adenosine diphosphate to provide immediate energy for contractions (ATP -> ADP + P + energy). Because the amount of stored ATP is small, it must be replaced with the splitting of PCr. Oxidation of carbohydrate and fat replenish the limited stores of ATP and PCr.

Think of a muscle as a controlled combustion chamber where the energy stored in carbohydrate and fat is slowly transferred to ATP. The key to the process is enzymes, organic catalysts that release and transfer energy.

1. Nerve impulse triggers splitting of ATP to provide energy.
2. PCr splits to provide energy to resynthesize ATP.
3. Glucose (or glycogen) is broken down (glycolysis) to lactic acid and ATP.
4. Glucose, glycogen, or fat are oxidized in muscle mitochondria to form carbon dioxide, water, and energy to form ATP (see figure 11.3).

Steps 1, 2, and 3 are nonoxidative, or anaerobic; they do not require the presence of oxygen. Step 4 requires oxygen, so it is called aerobic. Anaerobic metabolism of glucose leads to the formation of two molecules of ATP, whereas the aerobic or oxidative metabolism of glucose yields 38 molecules of ATP. Aerobic metabolism is a far more efficient use of fuel than anaerobic metabolism.

Available Energy

ATP and PCr are good for 3 to 4 calories of energy and can be exhausted in a few seconds of maximal effort, such as running up stairs. As ATP and PCr are depleted, anaerobic glycolysis kicks in to produce more ATP from the nonoxidative breakdown of glucose. But because the anaerobic pathway is not efficient and muscle glycogen stores are limited, the aerobic, or oxidative, breakdown of carbohydrate and fat increases to provide a long-term supply of ATP. After an hour or more of vigorous effort, the muscle glycogen supply begins to decline. The muscle then gets its glucose from the blood. Blood glucose comes from liver glycogen. When that supply is depleted, blood glucose declines and fatigue is imminent. Muscle cannot work as hard without carbohydrate, and the brain and nervous system require carbohydrate for energy (see figure 11.4).

Energy Use

For prolonged, arduous work, we must use aerobic pathways. They are more efficient than anaerobic pathways, and the fuels are more abundant. Table

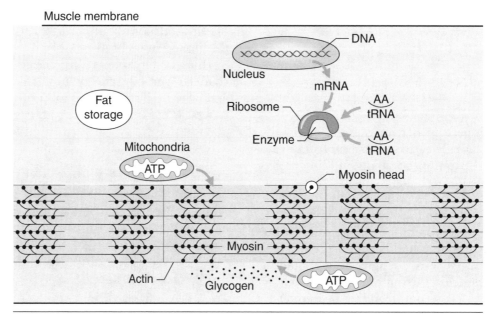

Figure 11.3 Basic structures in a muscle fiber.

Reprinted, by permission, from B.J. Sharkey and S.E. Gaskill, 2007, *Fitness and health*, 6th ed. (Champaign, IL: Human Kinetics), 87.

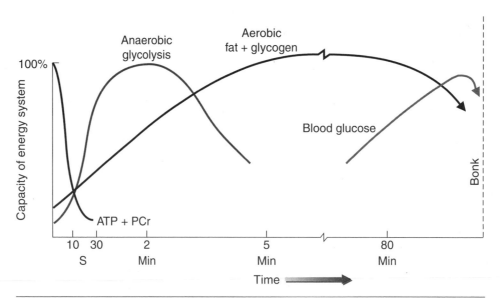

Figure 11.4 Pattern of energy use. As muscle glycogen is used up, blood glucose temporarily fills the demand for carbohydrate.

Reprinted, by permission, from B.J. Sharkey and S.E. Gaskill, 2007, *Fitness and health*, 6th ed. (Champaign, IL: Human Kinetics), 213.

Table 11.2 Available Energy Sources

Source	Supply	Energy (kcal)	Miles*
ATP and PCr	Small amount in muscles	4-5 kcal	0.045
CARBOHYDRATE			
Muscle glycogen	20 g per kg muscle	1,600	16
Liver glycogen	80 g	320	3.2
Blood glucose	4 g	16	0.16
FAT			
Muscles	Limited, varies with training	1,500	15
Adipose tissue	Variable**	30,000-70,000	300-700

*Assume 100 kcal/mile and all energy in working muscles.

**Depends on body weight and % body fat: 10% fat × 150 lb = 15 lb fat × 3,500 kcal/lb = 52,500 kcal.

PCr = phosphocreatine

Adapted, by permission, from B.J. Sharkey and S.E. Gaskill, 2006, *Sport physiology for coaches* (Champaign, IL: Human Kinetics), 126.

11.2 portrays the supply of available energy. Fat is, by far, the most abundant source of available energy, supplying enough for someone to run hundreds of miles. Carbohydrate stores are more limited, but they can be raised with training and diet (carbohydrate loading) or supplemented with solid and liquid energy sources during work.

The contribution of fat and carbohydrate varies during work. For light work, muscles use mostly fat, supplied with muscle triglyceride and plasma free fatty acids (FFA) mobilized from adipose tissue and transported in the bloodstream. As exercise intensity increases, carbohydrate use increases, becoming predominant at high levels of effort (see figure 11.5).

A small amount of fat is stored in muscle. This muscle triglyceride may be increased with endurance training. During long-duration effort, fat use increases with time. Fat mobilization from adipose tissue is delayed in the first half-hour of exercise. But as the activity continues, fat use increases. Training increases fat use, with early supplies coming from muscle triglyceride (Holloszy et al. 1986). As training increases fat use, it spares the limited supplies of muscle and liver glycogen, an important factor in work or sport.

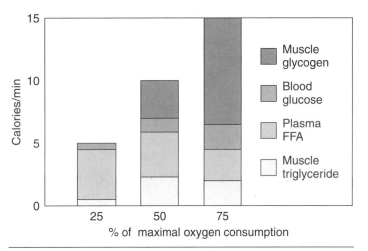

Figure 11.5 Energy sources and exercise intensity. Approximate contribution of energy sources at three levels of exercise. Plasma free fatty acids (FFA) are transported via the blood to the muscles.

Reprinted, by permission, from B.J. Sharkey and S.E. Gaskill, 2007, *Fitness and health*, 6th ed. (Champaign, IL: Human Kinetics), 233.

Oxygen and Energy

Oxygen is the key to prolonged activity. The ability to take in, transport, and use oxygen in muscles, aerobic fitness, is essential for the performance of physically demanding occupations. When we can't supply sufficient oxygen to the muscles, we are forced to use inefficient anaerobic pathways and limited sources of energy, such as ATP, PCr, and glycogen. When we begin to exercise, oxygen intake does not immediately meet the demand. An oxygen deficit results as we rely on ATP, PCr, and anaerobic glycolysis (leading to the formation of lactic acid). When oxygen intake begins to meet the demand, a steady state is achieved and exercise can continue so long as we are able to meet the fuel and oxygen requirements. After exercise, oxygen returns slowly to resting levels. Postexercise oxygen intake in excess of resting needs is called the oxygen debt. The debt, or excess postexercise oxygen consumption (EPOC), is used to repay the oxygen deficit, replace ATP and PCr, remove lactic acid, and replace some of the energy used during the exercise (see figure 11.6).

Now let's learn how the respiratory and cardiovascular systems carry oxygen and energy to working muscles.

Supply and Support Systems

The respiratory and cardiovascular systems, and to a degree the endocrine system, function as supply and support systems for working muscles. The respiratory system takes air into the lungs and allows the transport of oxygen to red blood cells in the bloodstream. The cardiovascular system transports blood to the heart and then to the muscles, where oxygen and energy are used. The muscles produce by-products such as carbon dioxide, lactic acid, and heat, which are transported from the muscle. Carbon dioxide is eliminated during exhalation, lactic acid is buffered or used as a fuel by other muscle fibers, and heat is dissipated via the evaporation of sweat. The hormones of the endocrine system also play a part in supporting muscles.

Respiratory System

Respiration has two main functions: getting oxygen into the body and getting rid of carbon dioxide. Ventilation (V), the amount of air you breathe per minute, is the product of respiratory rate (or frequency or f) and the volume of air per breath (tidal volume or TV), or $V = f \times TV$. Ventilation during hard work ranges from 40 to 60 liters per minute ($40 L = 20 \times 2L$), well below the ventilation of maximal effort ($V = 120 L/min = 40 \times 3L$). Because fit workers have a lower frequency and larger tidal volume, their respiration is more efficient. The ventilation during a maximal oxygen intake test is well below the maximal ventilatory volume (MVV = 180 L/min). The healthy respiratory system is overbuilt for its job. Unfortunately, age reduces that margin, especially in those who elect to remain sedentary.

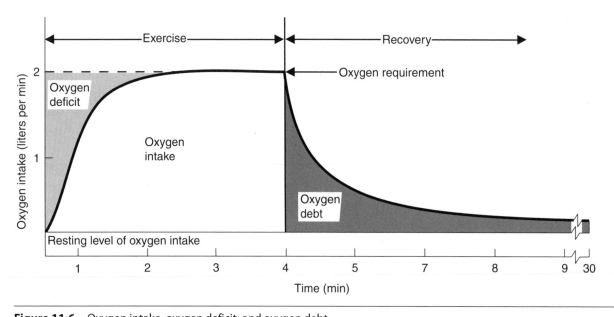

Figure 11.6 Oxygen intake, oxygen deficit, and oxygen debt.

Reprinted from B.J. Sharkey, 1990, *Physiology of fitness*, 3rd ed. (Champaign, IL: Human Kinetics), 281.

When the diaphragm contracts, it creates an area of lower pressure, causing air to rush into the lungs. When air reaches the tiny air sacs called alveoli, oxygen crosses the alveolar and capillary membranes and hitches a ride on hemoglobin attached to red blood cells (see figure 11.7). Under favorable conditions, hemoglobin saturation with oxygen is about 97 percent of the available space. As one goes up in altitude and the barometric pressure falls, the hemoglobin saturation declines. The respiratory system joins with the kidneys and several different blood-borne buffers to control the acid–base balance, which can be disturbed with acid by-products during very vigorous effort, especially in untrained workers.

Cardiovascular System

We have followed oxygen from the atmosphere to the blood. Now let us see how it gets to the muscles. The blood serves to transport oxygen and carbon dioxide as well as fuels, waste products, hormones, antibodies, and heat. Blood buffers (hemoglobin, buffer systems, and protein molecules) help to regulate the acid–base balance. Our 5 liters of blood contain about 5 million red cells per cubic millimeter. The hemoglobin on red cells uses iron to carry oxygen, and the healthy diet

must include sufficient iron to avoid anemia. One important outcome of endurance training is a 10 to 15 percent increase in blood volume, which aids oxygen delivery and increases the ability of workers such as firefighters and iron workers to tolerate hot working conditions.

The heart is the ultimate endurance muscle, amply supplied with mitochondria for oxygen use. It has a well-developed system of blood vessels (coronary arteries) for delivery of oxygen and fuel to cardiac muscle. The heart consists of two pumps: the right side, which sends blood to the lungs, and the left side, which pumps blood through the rest of the body (see figure 11.8).

The output of the cardiac pump depends on two factors: the rate of the pump (heart rate, HR) and the volume per stroke (stroke volume, SV); cardiac output = HR × SV. With a resting HR of 72 beats per minute and a stroke volume of 70 ml, the cardiac output is about 5 liters of blood per minute. Heart rate increases with work rate ($\dot{V}O_2$).

Figure 11.9a illustrates the rise in heart rate for a trained worker and an untrained worker. Figure 11.9b shows how stroke volume responds for the workers. Clearly, the trained worker has a larger stroke volume, allowing a much larger cardiac output and a much greater supply of oxygen to the muscles during exertion.

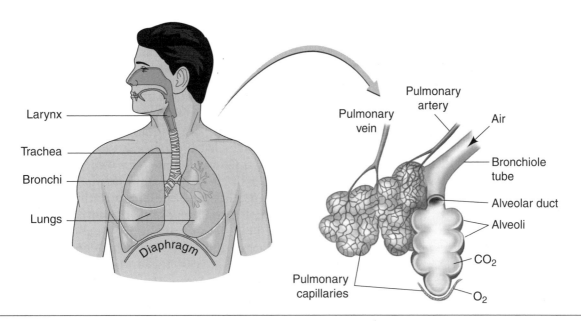

Figure 11.7 Respiratory system.

Reprinted, by permission, from B.J. Sharkey and S.E. Gaskill, 2007, *Fitness and health*, 6th ed. (Champaign, IL: Human Kinetics), 89.

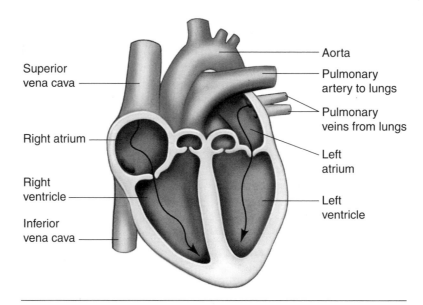

Figure 11.8 The heart.

Reprinted, by permission, from B.J. Sharkey and S.E. Gaskill, 2007, *Fitness and health*, 6th ed. (Champaign, IL: Human Kinetics), 91.

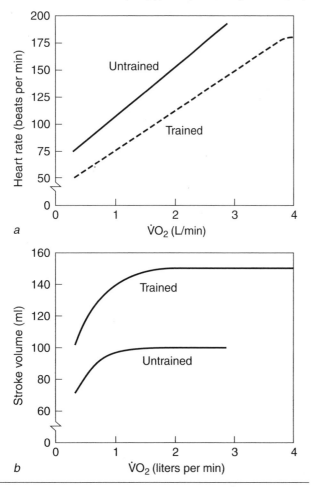

Figure 11.9 Relationship of oxygen intake (V̇O₂) to heart rate (a) and stroke volume (b). At workloads above 1.5 L/min, any increase in the cardiac output results from an increase in the heart rate and not stroke volume, which tends to plateau.

Adapted, by permission, from B.J. Sharkey and S.E. Gaskill, 2006, *Sport physiology for coaches* (Champaign, IL: Human Kinetics), 134.

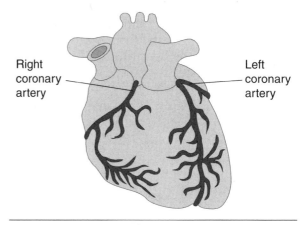

Figure 11.10 Coronary arteries. The left coronary artery branches to serve the muscular left ventricle.

Reprinted from B.J. Sharkey, 1990, *Physiology of fitness*, 3rd ed. (Champaign, IL: Human Kinetics), 288.

The coronary arteries receive blood when it leaves the left side of the heart (see figure 11.10). When the arteries are narrowed with atherosclerosis, the heart is compromised and at risk for a heart attack (myocardial infarction). For sedentary people the risk rises over 50 times during exertion. Aerobic fitness training lowers the risk of heart disease from 30 to 70 percent, depending on the degree of activity. Those who acquire as many as 3,500 kilocalories of exercise per week achieve the greatest reduction (see figure 4.1, page 39). For example, if you burn 100 calories walking or running a mile, you would have to do five miles a day, seven days a week to achieve the maximum reduction in cardiovascular risk (see p. 119).

IIIIII **WHAT IS A CALORIE?** IIIIII

A calorie (technically, a kilocalorie) is a unit of energy defined as the amount of heat required to raise the temperature of 1 kilogram of water 1 degree Celsius. We store calories when we eat and burn them when we exercise. Body weight influences caloric expenditure during exercise. A 180-pound (81.6 kg) person burns more calories running at a certain pace than one who weighs 150 pounds (68 kg). The first burns 136 calories per mile (85 calories per kilometer), whereas the second burns 113 calories per mile (70 calories per kilometer).

Endocrine System

Hormones of the endocrine system support muscle function. Several hormones can increase blood glucose levels (glucagons, epinephrine, norepinephrine, and cortisol), but insulin is the only one that can move glucose into the muscle. Fortunately, glucose is able to enter muscles during physical activity, even in the absence of insulin. That is one reason physical activity and diet are important components of diabetes treatment. Physical training improves insulin sensitivity, reducing the amount of insulin required. Hormones are also involved in the mobilization of fat from adipose tissue during exercise and the conservation of body water during work in the heat. Growth hormone promotes the growth of bone and muscle and the use of fat for energy (thereby sparing glycogen and glucose). Other hormones, such as adrenaline (epinephrine) and cortisol, are involved during vigorous effort and stress.

Training Effect

When we engage in exercise such as jogging or lifting weights at a level above our normal daily activity (load), we overload the muscles and their support systems, including the heart and lungs. When we repeat the exercise regularly (e.g., every other day), our bodies begin to adapt to the overload imposed by the exercise. We call the adaptations the training effect. How does exercise signal muscle fibers and support systems to undergo changes that will permit more exercise in the future? The answer lies in the genes and DNA.

Genes influence potential, but they don't ensure it. The 30,000 genes that form the blueprint of the human body are subject to the influence of the environment and behavior. Our genotype is our genetic constitution, whereas our phenotype is the observable appearance resulting from the interaction of the genotype and the environment. In sport, genetic potential can only be realized when genes are switched on via the process of training. Genes carry the code for the formation of proteins. A specific type of training turns on a promoter that activates specific genes. Through a process called transcription, an RNA strand is formed on the template made up of the gene's DNA. The RNA becomes the messenger (mRNA) that exits the nucleus, enters the cytoplasm, and binds to a ribosome, whose function is to synthesize protein (see figure 11.3 on page 114). The mRNA translates the genetic code into a sequence of amino acids to form a specific protein. Transfer RNA (tRNA) reads the mRNA blueprint and then captures appropriate amino acids for use in the synthesis of the desired protein. Without training, genetically gifted people cannot achieve success in work or sport. But heredity is much more complicated than genes, DNA, and RNA.

The genetic response to a specific form of training resides in a cluster of genes, and each can be influenced by the presence or absence of so-called enhancers, regulatory elements that influence the degree of response. That helps to explain why the individual response to identical training is so variable. Factors that may influence the response to training could include maturation (hormones), nutrition (energy, amino acids), adequate rest, and even chemically related emotional factors such as stress, which mobilizes epinephrine (adrenaline) and cortisol.

Endurance training leads to an increase in the concentration of oxidative enzymes and to a rise in the size and number of mitochondria, the cellular power plants where all oxidative metabolism takes place (Hood et al. 2000). These particular adaptations are specific to endurance training, and they take place only in the muscles used in training. Endurance training also improves the function of respiratory muscles and the heart, and it increases the blood volume, which improves stroke volume and cardiac output. The rise in blood volume helps explain why the heart rate declines with training.

With a greater blood volume and blood redistributed from other regions (e.g., the digestive system), more blood enters the heart and is pumped out with each beat. This allows us to do the same work with a lower heart rate (and larger stroke volume). A large portion of the enhanced cardiorespiratory function of endurance athletes is due to the training-induced increase in blood volume and its effect on cardiac function, increased hemoglobin and stroke volume, and decreased heart rate and blood viscosity (Gledhill, Warburton, and Jamnik 1999). Training-induced changes in blood volume accounted for almost half of the improvements in $\dot{V}O_2$max with training: The results are similar for continuous or interval training (Warburton et al. 2004) (see below).

⁞⁞⁞⁞⁞ HEART RATE ⁞⁞⁞⁞⁞ FOR TRAINING

For many people, determining the training heart rate is a needless distraction from the pleasures of exercise. If you enjoy the discipline of heart rate measurements, however, continue to use them. You can use a percentage of your maximal heart rate (e.g., HRmax = 220 – age) or the heart rate range formula (range = HRmax – resting HR) to identify your training heart rate. The heart rate range formula calculates a heart rate equal to the percentage of $\dot{V}O_2$max. For a heart rate equivalent to 70 percent of $\dot{V}O_2$max for a 50-year-old (220 – 50 = 170):

$$HR = [70\% \times (HRmax - resting\ HR)]$$
$$+ resting\ HR$$
$$= [70\% \times (170 - 70)] + 70$$
$$= 140\ bpm$$

By contrast, 70 percent of the maximal heart rate (170) equals 119 beats per minute, which is approximately equal to 55 percent of $\dot{V}O_2$max. The heart rate range is sometimes used to adjust for measured differences in the resting and maximal heart rates and to avoid errors in the estimation of the training heart rate.

Strength training leads to the production of contractile proteins, actin and myosin. Other specific effects of strength training include an increase in muscle mass and a toughening of connective tissue. With training we learn to exert force more effectively. Core training to strengthen back and abdominal muscles contributes to performance and a lower risk of musculoskeletal injury (see chapter 12). Because aerobic and muscular training lead to the production of proteins, both forms of training require a diet adequate in energy and protein.

Summary

This chapter outlined the structure and function of muscle fibers and the energy sources used for contractions. We demonstrated that fat and carbohydrate are the primary sources of energy for prolonged hard work, and that we turn to carbohydrate for vigorous effort. We dealt with the importance of oxygen in energy metabolism and the roles of the respiratory, cardiovascular, and endocrine systems. Finally, we described the mechanism by which training improves the function of muscles and support systems.

Training is specific: When we train for aerobic fitness, we synthesize enzyme protein that improves our ability to use oxygen; when we train for strength, we synthesize contractile proteins that improve muscle force. Training takes place in the muscles used in training. We cannot train arm muscles to improve leg endurance, or vice versa. We must train the muscles we will need on the job. Job-related fitness training is the subject of the next chapter.

For a biochemical evaluation of workplace stress, see appendix E, page 207. For more information on the physiology of work, consult Astrand and colleagues' (2003) *Textbook of Work Physiology* or Wilmore and Costill's (2006) *Physiology of Sport and Exercise*, both published by Human Kinetics.

Chapter 12
Job-Related Fitness

© Bananastock

The majority of jobs available to personnel entering the U.S. military services are physically demanding, and soldiers must maintain high levels of physical fitness to perform their duties optimally. In nonmilitary physically demanding occupations, the courts seem interested in establishing a minimally acceptable performance level for recruits. After recruits are hired, they are expected to enhance their job-related knowledge, skills, and abilities (KSAs). However, few organizations require current workers to demonstrate maintenance of a minimum level of job-related physical fitness. This chapter describes the development of a physical maintenance program required of all employees, but especially those hired at the minimum acceptable performance level.

This chapter will help you do the following:

- Develop a job-related fitness program.
- Understand the dimensions of aerobic fitness and measure sustainable fitness.
- Differentiate among components on the strength–endurance continuum.
- Select exercises for core training.
- Understand why periodization is important for muscular and aerobic training.
- Determine the job-related elements of body composition.
- Identify the major issues associated with a job-related fitness program.
- Calculate the risks and benefits of job-related fitness programs.

Job-Related Fitness Programs

As athletes strive to improve performance through effective training techniques, so too can workers benefit from optimally designed exercise training programs designed to maintain and enhance occupational physical performance. However, few organizations require workers to maintain a minimum level of job-related physical fitness, let alone an enhanced level of work capacity. Over a typical 25-year career, the result is a significant decline in job-related fitness (see figure 12.1).

Figure 12.2 indicates the hiring level required to ensure career-long work capacity in the absence of a job-related fitness training program. We assume a decline in fitness of 1 percent per year, or 10

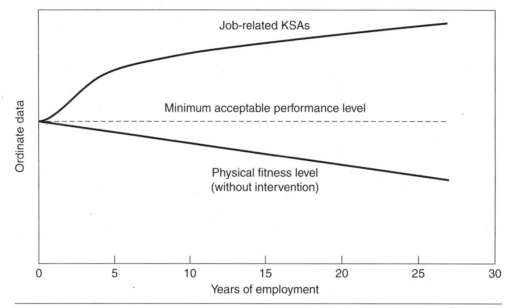

Figure 12.1 Decline in fitness (without intervention). KSAs = knowledge, skills, and abilities.

Davis 2007 (appendix A).

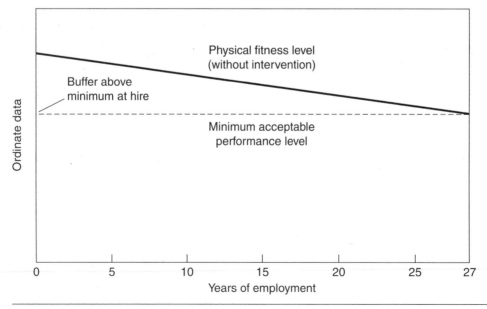

Figure 12.2 Hiring level to ensure acceptable performance throughout career.

Davis, 2007 (appendix A).

percent per decade. The sloping line indicates the rate of decline for a person who remains active (5 percent decline per decade). This is the only way to ensure performance in the absence of training or work standards. Unfortunately, the courts have not approved this approach to the maintenance of a minimum performance level.

In a document titled *Physical Training for Improved Occupational Performance*, the American College of Sports Medicine (2002) noted that job performance in physically demanding occupations can be improved and made safer by having workers participate in fitness programs targeted at improving aerobic and muscular fitness. Tasks that involve prolonged and repetitive pulling, pushing, holding, carrying, and lifting can lead to cumulative trauma disorders such as low back pain, sprains, strains, and neck pain. Physical training can be used as a preventive or rehabilitative tool in occupational settings, while also increasing worker productivity by improving aerobic capacity, muscle strength, endurance, or power. Training can also prevent mismatches between physical capabilities and job demands and decrease incidences of injury-related absenteeism.

Program Development

Because the job-related fitness program is designed to enhance job performance, it's a good idea to plan the program with the help of the job task analysis (discussed in chapter 4) that identified physically demanding tasks that are critical to performance. Employers may decide to engage the services of an exercise physiologist who has studied the job and knows a lot about the workers. The physiologist can select exercises, design training, and make recommendations concerning equipment purchases.

Once the important tasks have been identified, program developers will need information regarding the size of loads, distances moved, and frequency and duration of task performance. In addition to the lifting guidelines in chapter 15, program developers will need additional information on the positions of loads, body positions, and movements. The tasks must be analyzed to determine the energy systems involved, muscle groups employed, requirements for strength or endurance, and environmental conditions during work. This information can determine the parameters of the job-related fitness program.

Job-Related Training

Training designed to improve performance in sport or work must include the physically demanding tasks of the activity. If the job requires an optimal level of aerobic fitness, the training should develop that component in a manner specifically related to the job. If the job demands lifting heavy boxes, the training should strengthen the muscle groups used in the task. Most demanding jobs require aerobic and muscular fitness. Training begins at a level tolerable to the worker and progresses gradually to provide an overload that stimulates improvements in performance. Job-related training has enhanced the work performance of men and women in a variety of tasks.

‖‖‖‖‖‖‖ **READY** ‖‖‖‖‖‖‖
FOR COMBAT

In the U.S. military, technology is increasing the physical demands of some jobs and reducing the demands of others. As a result, the physical demands of some military specialties could be lower than those of others. However, it is Department of Defense policy that every uniformed service member must be combat ready and that physical fitness is essential to combat readiness and is an important part of the general health and well-being of armed forces personnel (Sackett and Mavor 2006).

Aerobic Fitness

In chapter 2 we identified two ways to quantify aerobic fitness: aerobic capacity (L/min) and aerobic power (ml $\cdot$ kg^{-1} $\cdot$ min^{-1}). We noted that aerobic power is a better measure of performance when workers are carrying their own body weight, as in walking and climbing. The performance of heavily loaded soldiers doing a 2-mile (3.2 km) run/walk with a 70-pound (31.8 kg) pack is related ($r = .75$) to aerobic capacity ($\dot{V}O_2$max in L/min), whereas the correlation between the unloaded run/walk and aerobic capacity is lower. The unloaded run/walk time for 2 miles (3.2 km) correlated ($r = .74$) to aerobic power (ml $\cdot$ kg^{-1} $\cdot$ min^{-1}) and somewhat less ($r = .67$) to aerobic capacity (Gutekunst, Harmen, and Frykman 2005). Now we are going

Performance of heavily loaded soldiers doing a 2-mile (3.2 km) run/walk with a 70-pound (31.8 kg) pack is related ($r = .75$) to aerobic capacity ($\dot{V}O_2$max in L/min).

Lactate Thresholds

Lactic acid is both an energy carrier and a metabolic by-product of intense effort. Its accumulation is a sign that you are using energy faster than you can produce it aerobically. Too much lactic acid interferes with the muscle's contractile and metabolic capabilities. Lactic acid and the high levels of carbon dioxide produced in vigorous effort are associated with labored breathing, fatigue, and discomfort. Blood lactate is a by-product of anaerobic glycolysis.

In a progressive treadmill test, going from a walk to a jog to a run, lactate rises slowly at first and then more rapidly (see figure 12.3). The transition from slow oxidative muscle fibers to fast glycolytic fibers is associated with the increase in lactic acid. The first lactate threshold (LT1) occurs at about 50 percent of $\dot{V}O_2$max. The first threshold defines a level of exertion that can be continued for several hours to daylong, depending on its level. As we go from a jog to a run, we involve additional fast-twitch muscle fibers and produce more lactate. The second lactate threshold (LT2) defines the upper reaches of aerobic metabolism. It is highly correlated to performance in competitive events (e.g., a 10K run) lasting from 30 minutes to 3 hours (Sharkey and Gaskill 2006).

Training can raise $\dot{V}O_2$max and both lactate thresholds. Eventually the $\dot{V}O_2$max will plateau, but the thresholds continue to improve (see figure 12.4). As training improves the muscle's oxidative capacity, more work can be done without an increase in lactic acid. The first lactate threshold defines the ability to perform prolonged arduous work. People with an improved LT1 can sustain 50 percent or more of their $\dot{V}O_2$max for eight or more hours. If a worker has a $\dot{V}O_2$ max of 50 ml $\cdot$ kg^{-1} $\cdot$ min^{-1}, and his LT1 is 50 percent of his max, his $\dot{V}O_2$ at LT1 is 25 ml $\cdot$ kg^{-1} $\cdot$ min^{-1}.

Studies in the workplace indicate that LT1 defines work output in some physically demanding occupations. Wildland firefighters were divided into two groups, those with higher and lower levels of LT1. The firefighters were fitted with electronic activity monitors to determine work activity and sent out to work for nine days on an actual fire.

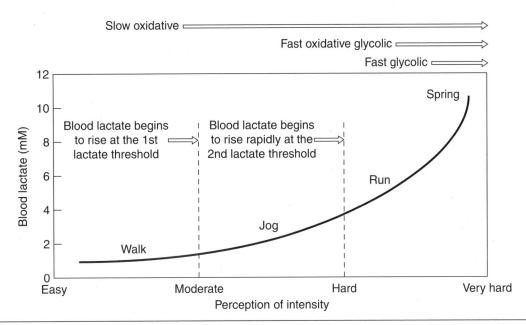

Figure 12.3 Lactate thresholds. As exercise intensity (percentage of $\dot{V}O_2$max) increases, we recruit additional slow oxidative (SO) fibers, fast oxidative glycolytic (FOG) fibers, and finally fast glycolytic (FG) fibers. The first lactate threshold (LT1) occurs as muscles produce more lactic acid than can be cleared from the blood. This generally occurs at a moderate intensity and represents long-duration, sustainable capacity. The second lactate threshold (LT2) occurs when blood lactate accumulates rapidly and breathing becomes labored. Exercise at LT2 cannot be sustained for long.

Reprinted, by permission, from B.J. Sharkey and S.E. Gaskill, 2007, *Fitness and health*, 6th ed. (Champaign, IL: Human Kinetics), 73.

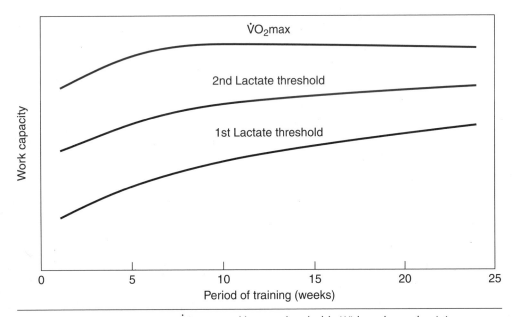

Figure 12.4 Improvements in $\dot{V}O_2$max and lactate thresholds. With prolonged training, $\dot{V}O_2$max (aerobic fitness) begins to plateau, but the capacity to perform submaximal work at the first and second lactate thresholds continues to improve.

Reprinted, by permission, from B.J. Sharkey and S.E. Gaskill, 2007, *Fitness and health*, 6th ed. (Champaign, IL: Human Kinetics), 111.

Those with higher levels of LT1 (43.7 ml · kg⁻¹ · min⁻¹) performed more work throughout the day than did those with lower values (34.6 ml · kg⁻¹ · min⁻¹) (Gaskill et al. 2002). They were able to sustain a higher work rate with no greater sense of fatigue (see figure 12.5).

Army Fitness Training Units

A recent analysis of the success of first-term U.S. soldiers concluded that people who enter the army in poor physical condition are less likely to complete their initial training (Buddin 2005). New recruits are assigned to fitness training units (FTUs) if they fail a fitness test that is administered at the beginning of training. The course is intended to prepare new recruits for the demands of basic combat training and reduce injuries during training. Unfortunately, the FTU training is doing little to counter the tendency of participants to struggle in the army. Surprisingly, FTU participants who leave the army do so for performance and conduct reasons, not specifically for fitness, suggesting that FTU participants may enter the army with problems in addition to fitness. It costs the U.S. Army more than $15,000 to recruit one soldier, and the army must recruit 80,000 to 90,000 each year. When a soldier doesn't complete the first year, the army must spend the same amount to find a replacement. Poor fitness can be expensive (www.rand.org).

Sustainable Fitness

Dr. Steve Gaskill, from the University of Montana Human Performance Laboratory, has labeled the $\dot{V}O_2$ at LT1 "sustainable fitness." It defines the workload one can sustain throughout the working day. During the nine-day study (see figure 12.5) those in the high-fit group averaged slightly more work than their oxygen intake at LT1. Those in the lower-fit group worked somewhat above their sustainable fitness level. Sustainable fitness (LT1) can be trained by gradually increasing the duration of time spent at or somewhat above LT1. Of course, since training is specific, much of the training should involve the work itself or a close approximation of the work.

Sustainable fitness can be measured by requiring sustained work for a sufficient duration. For example, wildland firefighters must pass the pack test, a 3-mile (4.8 km) hike with a 45-pound (20.4 kg) pack, within 45 minutes. The energy cost of the test is the same as that of the job, so firefighters are asked to demonstrate the ability to sustain that workload for the time it takes to complete the hike. Those who complete the test in less than 45 minutes demonstrate the ability to sustain a higher work rate. Completing the test in 45 minutes predicts a $\dot{V}O_2$max of 45 ml · kg^{-1} · min^{-1}, another dimension of aerobic fitness (Sharkey 1991).

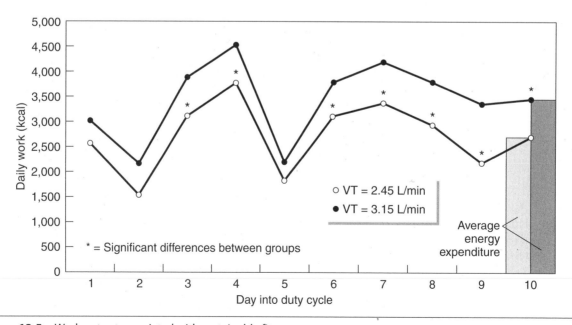

Figure 12.5 Work output associated with sustainable fitness.

Data from S.E. Gaskill et al., 2002, "Fitness, work rates, and fatigue during arduous wildfire suppression," *Medicine and Science in Sports and Exercise* 34: s195.

To train for sustainable fitness, workers should engage in work-related training several days a week. They should go long and slow on one day, go shorter and faster on the second day, and do medium distance and pace on the third day. This aerobic training should be done every other day, with muscular fitness training on alternate days. They should add job-related cross-training one or two days per week. For example, wildland firefighters train to carry loads up and down hills. They can cross-train with mountain biking, load packing, or trail running. When the fire season draws near, the crew boss engages crew members in project work to continue the training and to accomplish work hardening.

Workers should begin training for LT1 three times per week. After several weeks they should add another day of training and introduce LT2 training and some related cross-training. After six to eight weeks they could try some high-intensity training for variety. But unless the job requires very high intensity, the $\dot{V}O_2$max training isn't necessary. The goal of job-related training is to improve work performance via sustainable fitness. Continued LT1 training will bring about improvements in sustainable fitness for many months, even as the $\dot{V}O_2$max plateaus. For much more detail on training, consult *Sport Physiology for Coaches* (Sharkey and Gaskill 2006).

In table 12.1 we introduce the rating of perceived exertion (RPE) scale as a way to gauge training intensity. The scale, developed and validated by Dr. Gunnar Borg, is correlated to exercise intensity and closely related to heart rate (see table 12.2).

Wildland firefighters train to carry loads up and down hills. They can cross-train with mountain biking, load packing, or trail running.

Photo courtesy of the USDA Forest Service, Missoula Technology and Development Center

Table 12.1 Aerobic Training

Component	Goal	Intensity*	Duration	Frequency
LT1	Sustainable fitness	RPE 11-12	Long (hr)	3-5x/wk**
LT2	Short/intense	RPE 15	3-5 min	1-2x/wk
$\dot{V}O_2$max	High intensity	RPE 16	2 min	1x/wk***

*RPE = rating of perceived exertion (see table 12.2).
**Plus some related cross-training.
***Add after 6-8 weeks of training.

Table 12.2 Rating of Perceived Exertion

Scale Descriptor	Borg Scale
Very, very light	6
	7
	8
Very light	9
	10
Fairly light	11
	12
Somewhat hard	13
	14
Hard	15
	16
Very hard	17
	18
Very, very hard	19

People should exercise at the level indicated in table 12.1. When they work at a given level (e.g., 15 on the scale), their heart rate approximates the number times 10 (15 × 10 = 150). The RPE scale, an excellent way to gauge training, adjusts to environmental factors (heat and humidity), fatigue, and illness.

Muscular Fitness

The job task analysis indicates the muscle groups involved and the type of training they require. Employees whose strength is low will require strength training. If strength is adequate but endurance is lacking, that should be the goal of training. If some tasks require power, that too can be improved.

In training we often speak of the overload principle, which states the following:

- For improvements to take place, workloads have to impose a demand (overload) on the body system (above two-thirds of maximal force for strength).

- As adaptation to loading takes place, more load must be added.

- Improvements are related to the intensity (tension for strength), duration (repetitions), and frequency of training.

Overload training leads to adaptations in the muscles according to the type of training. Here again, the principle of specificity applies, as it does with aerobic training. The adaptation to strength training includes increased size due to increases in contractile proteins (actin and myosin) and tougher connective tissue. These and adaptations in the nervous system allow the muscle to exert more force. The specific adaptations to muscular endurance training include improved aerobic enzyme systems, larger and more numerous mitochondria (increased mitochondrial density), and more capillaries. All these changes promote oxygen delivery and utilization within the muscle fiber, thereby improving endurance. Fatiguing repetitions stimulate the muscle to become better adapted to use oxygen and aerobic enzymes for the production of energy. When we perform many repetitions, we become better able to use fat as a source of energy.

Table 12.3 reviews the effects of each type of training on the strength–endurance continuum. It shows that high-resistance training increases strength and that low-resistance repetitions increase muscular endurance. Training that falls between high resistance and low resistance leads to the development of short-term (anaerobic), or intermediate, muscular endurance. The table can be used to identify training goals.

Trainees should begin strength training with one set per day for each of 8 to 10 exercises. They should then add another set each week until they are doing three. After several months of training they could add another set. They should do three or four sets of 6- to 10-repetitions maximum (RM), three times per week, allowing at least two minutes between sets. Trainees should do medium resistance (8RM) one day per week, high resistance (6RM) at least one day per week, and 10RM one day per week.

When they can do three sets with all the repetitions (e.g., 6RM), it is time to increase the load.

Trainees should follow the prescriptions for short, intermediate, or long-term endurance. For power, they should use the short-term endurance prescription (three sets with 15 to 25RM), using 30 to 60 percent of maximal resistance *as fast as possible*. When strength and endurance are adequate for the job, trainees should switch to a two-days-per-week maintenance program.

Photo by Lance Cpl. Darhonda V. Hall, U.S. Marines

Overload training leads to adaptations in the muscles according to the type of training.

Table 12.3 The Strength–Endurance Continuum

	Strength	Endurance	
		SHORT-TERM	LONG-TERM
Develop	Maximum force	Brief (2-3 min) persistence with heavy load	Persistence with lighter load
Prescription	8-12RM 3 sets	15-25RM 3 sets	30-50+RM 2 sets
Improves	Contractile protein	Some strength	Aerobic enzymes
	Connective tissue	Anaerobic metabolism	Mitochondria

RM = repetitions maximum

Reprinted, by permission, from B.J. Sharkey and S.E. Gaskill, 2007, *Fitness and health*, 6th ed. (Champaign, IL: Human Kinetics), 153.

Heli-Guiding Skill and Performance

In an investigation of helicopter ski guides, it was assumed that leg power and aerobic endurance would be correlated to performance because the job typically consists of 14 consecutive days of skiing approximately 21,325 vertical feet (6,500 vertical meters) per day. However, testing of 65 guides (age = 44 ± 7 years, height = 179 ± 8 cm, weight = 79 ± 11 kg, body fat = 17.9 ± 5.7 percent, with 14 ± 8 years of experience) revealed that the guides were so efficient in their skiing that the types of measures commonly used to evaluate skiing as a high-performance sport were not appropriate. The mean daily heart rate during a day of heli-guiding was found to be only 102 ± 13 beats per minute. Guides spent 118 ± 49 minutes at 45 to 55 percent $\dot{V}O_2$max, 17 ± 18 minutes between 56 and 60 percent, and 18 ± 29 minutes at 61 to 75 percent $\dot{V}O_2$max. Anaerobic leg power assessed by the vertical jump test was only 106 ± 5 kg/m/sec, and guides completed 44 ± 16 jumps onto a 16-inch (40.6 cm) bench in 90 seconds. In spite of these moderate scores on the fitness tests, the guides were able to ski hundreds of consecutive turns in varying snow conditions and load 5,060 pounds (2,300 kg) of skis on and off of the helicopter each day.

Contributed by Delia Roberts (2006).

Core Training

Everyone needs to maintain abdominal muscle tone and flexibility to avoid back problems. But those in physically demanding occupations, especially those involving lifting, need to go a step further. They need to engage in core training. The core muscles (abdomen, back, shoulders, hips), stabilize the primary movers used in lifting: the arms and legs. Core stability is part of the chain of power development in sport and work. This training is especially important when employees work in unusual positions, lift awkward loads, or move loads by rotating the trunk. Strong, well-conditioned trunk musculature is associated with reduced lumbar disc deformation and less stress on the lumbar spine (Debeliso et al. 2004). Core stability is related to strength, dynamic balance, and the balance among core muscle groups. Core training can improve performance and reduce low back problems, workers' compensation costs, and disabilities.

Abdominal training consists of exercises that employ upper and lower abdominal muscles (e.g., crunches, basket hang), with some twisting to engage the external obliques. Back muscles are best trained with a resistance that allows 15 repetitions (trunk lift, lateral lifts). Trainees should use light loads for trunk exercises and should not hyperextend. Shoulder and upper back muscles can be trained with weight machines or free weights. Trainees should gradually increase core training for eight weeks and then maintain core stability with one to two sessions per week. Flexibility exercises can be included for the back and hamstrings as part of the core program.

U.S. Navy photo by Photographer's Mate Airman Marvin E. Thompson Jr.

Sailors perform flutter kicks during physical training held at Nimitz Park on board U.S. Fleet Activities Sasebo, Japan. Core stability is part of the chain of power development in physically demanding work.

TRAINING GUIDELINES

As with all training, employers should use health screening or a medical exam to obtain medical approval for participation. All training sessions should begin with a warm-up, including stretching, and finish with a cool-down. Following are guidelines for muscular fitness training:

- Ease into the program with lighter weights and fewer sets. Begin with 15 to 25RM and gradually increase to 10RM.
- Avoid holding your breath during a lift.
- Exhale during the lift and inhale as you lower the weight.
- Always work with a companion or spotter when using free weights.
- Alternate muscle groups during a session; alternate arm and leg exercises. Allow two to three minutes of recovery time between sets of strength exercises.
- Keep records of progress. Test for maximum strength every few weeks. Also, record body weight and important dimensions (chest, waist, biceps). Vary the program; the fourth week should be somewhat easier than the preceding three to allow for recovery.

Periodizing the Training Plan

Periodization is the process of dividing training into sections or cycles, ensuring an adequate training stimulus as well as time for recovery. It is used for muscular and aerobic training. In muscular training for sport, periodization calls for cycles of strength, power, and endurance. Each cycle lasts from four to eight weeks, and each is divided into smaller cycles that provide for increasing loads and periods of relative rest. In work physiology, periodization could include eight weeks of strength training, four weeks of short-term endurance training, and eight weeks of long-term endurance training. During the strength cycle, the first week can include moderate loads (8RM), the second week higher loads (6RM), and the third week lighter loads (10RM) for recovery. Even the weekly program is periodized to allow increasing loads

CASE STUDY
Occupational Training

Although strength and aerobic training may not result in women attaining physical capacities equal to men, it may enable them to adequately perform many physically demanding military tasks. A study was conducted to determine whether women were capable of loading and firing a 155-millimeter howitzer (Sharp 1994). The 45-kilogram (99 lb) projectile had to be lifted to chest height and placed in the loading mechanism. The women received aerobic and muscular fitness training and were trained to load and fire the howitzer. After three weeks all women were capable of loading and firing the projectiles at the required rate, and some exceeded the standard.

(8RM, 6RM) followed by an easier day (10RM). Periodization provides a framework for gradual increases in training load with periods of relative rest and recovery. See *Sport Physiology for Coaches* (Sharkey and Gaskill 2006) for more on muscular fitness, core training, and periodization.

Body Composition

Body composition describes the fat and nonfat portions of the body. We can speak of percent body fat, or pounds of fat, and of lean body weight, or fat-free weight. We have already noted that lean body weight is significantly related to performance in lifting tasks. Those with more muscle can lift more. A weight training program will add pounds of muscle while paring away excess fat (in conjunction with aerobic training). Fat is a liability in the workplace; it burdens the worker; increases heat stress; and is a risk factor for diabetes, heart disease, and some cancers.

Weight Gain

Some workers lack sufficient lean body weight. They can improve performance by following a strength and weight gain program designed to add muscle, not fat. To gain muscle weight, workers engage in strength training while cutting back on aerobic training. They should eat adequate calories of energy and be sure to get sufficient quantities of high-quality protein. We recommend an extra 750 calories above energy expenditure on weight training days and an extra 250 calories on non-training days. This plan could lead to a weight gain of almost 1 pound (0.5 kg) per week. Trying to gain weight more quickly will result in much of it being in the form of fat.

Weight Loss

Most workers need to reduce body weight and fat. This requires a simple program of increasing caloric expenditure and reducing caloric intake. They can reduce caloric intake by cutting back on food portions and reducing high-fat and low-nutrient foods. In addition to the calories expended on the job, they can increase aerobic and muscular fitness training and do a long workout on the weekend. The goal for weight loss should never exceed 2 pounds (0.9 kg) per week. Obese people (BMI >30) may need to add behavioral therapy to the exercise and caloric restriction plan. For more on weight gain or loss, see *Fitness and Health* (Sharkey and Gaskill 2007).

Program Issues

Job-related fitness programs raise many questions: What standards should guide them; where should they be conducted; do employees train on company time; and what about people who can't or won't maintain work performance standards? We deal with program evaluation later in this chapter.

Program Standards

The American College of Sports Medicine (ACSM) is the worldwide authority on sports medicine and exercise science. ACSM publishes guidelines for exercise testing and prescription. ACSM also conducts the "gold standard" program for the certification of health and fitness instructors. We recommend using ACSM-certified instructors for health-related and job-related fitness programs (for more information, go to www.acsm.org).

Program Site

Many police and fire departments establish training facilities for their employees. Most large companies provide fitness facilities, even though few employees are engaged in physically demanding occupations. Some employers purchase the equipment with employee association funds. The alternative, for smaller organizations, is to contract with one or more health clubs, those with certified instructors. What matters most is that the training site is convenient, that the hours fit employees' schedules, and that the equipment meets the needs of the job-related program. A cost analysis will help employers decide which approach is best.

The training site should be convenient, and the equipment should meet the needs of the job-related program.

|||||||||||| FITNESS TRAIL ||||||||||||

Years ago the U.S. Forest Service developed a low-cost outdoor site at which to train and maintain fitness for wildland firefighters and others with demanding occupations (e.g., trail crews). The site consisted of a quarter-mile (0.4 km) running trail with 14 exercise stations. The exercise stations were constructed of logs, and the instructions were mounted on attractive signposts (see figure 12.6). Hundreds of these trails were constructed in areas where they could be shared with local communities, parks, and campgrounds. Although the fitness trail met a need years ago, before the proliferation of health and fitness clubs, few of the training sites remain in use today (Sharkey, Jukkala, and Herzberg 1984).

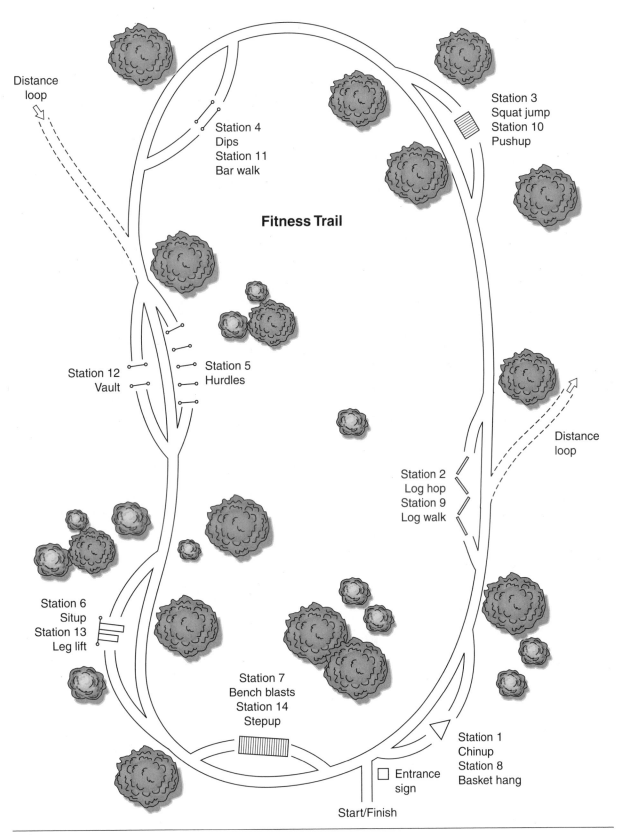

Distance loop

Station 4
Dips
Station 11
Bar walk

Station 3
Squat jump
Station 10
Pushup

Fitness Trail

Station 12
Vault

Station 5
Hurdles

Distance loop

Station 2
Log hop
Station 9
Log walk

Station 6
Situp
Station 13
Leg lift

Station 7
Bench blasts
Station 14
Stepup

Station 1
Chinup
Station 8
Basket hang

☐ Entrance sign

Start/Finish

Figure 12.6   Fitness trail.

From B.J. Sharkey, A. Jukkala, and R. Herzberg, 1984, *Fitness trail* (Missoula, MT: USDA Forest Service).

Company Time

Many fire and law enforcement organizations provide company time for fitness training. Some do a time-share, half company time and half employee time. Because neither of these arrangements is completely successful, we won't even bother to discuss placing the burden where it belongs, on the employee. Structural firefighters have ample time to train, but too few do so, even when motivated by economic reward. We recommend a time-share, with at least half of the responsibility borne by employees. Making substantial improvements could require over six hours per week, but fitness maintenance programs should not require much more than three hours per week.

Fitness Standards

In chapter 8 we discussed ways to deal with employees who do not pass periodic work capacity tests or meet fitness goals. The first option is to suspend field duties until they can meet the standards. That option may require more time for training and, if necessary, weight loss. When that commitment is not met, or if the employee cannot meet the standard after months of effort, it is time to consider reassignment. If alternative positions (dispatch, fire prevention, community relations, training) are not available, the employee may have to retire or be dismissed.

Risks and Benefits

This section considers the risks and benefits of the job-related fitness program and suggests costs and benefits that can be measured in the evaluation of the program. First, we address the risks of exertion and the risks that apply to the job, to fitness training, and to the periodic work capacity test. Then we discuss the benefits associated with a fitness program.

Exertion

Heart disease is the leading cause of death for both men and women. The major risk of exertion in a test, in training, or on the job is a significant cardiac event, such as a heart attack. Autopsy studies show, however, that heart disease begins early in life and develops at a rate that depends on heredity and

lifestyle (diet, physical inactivity, smoking, body weight). Although exertion may trigger a heart attack in a susceptible person (one with preexisting disease), it does not cause the disease. In fact, regular activity has been proven to substantially reduce the risk of heart disease and cardiac death. The American Heart Association listed physical inactivity as a major risk factor for heart disease (www.americanheart.org).

||||||||||||| **WARMING UP** ||||||||||||| **TO EXERTION**

Failure to warm up before vigorous exercise can result in electrocardiogram abnormalities, regardless of the fitness level or age of the subject. Dr. Barnard and associates found such abnormalities in 31 of 44 apparently healthy firefighters tested on a vigorous treadmill test (Barnard et al. 1972). The findings indicated inadequate blood flow in the coronary arteries and lack of oxygen to the heart. A warm-up consisting of a five-minute jog prevented the problem. It appears that athletes aren't the only ones who need to warm up; anyone performing vigorous work or exercise can benefit. Calisthenic warm-ups are common among factory workers in Europe and Japan, but they are less common in the United States. Though most workers can and should warm up, it is hard to see, for example, how law enforcement officers can do so before chasing a suspect. For this reason they should do all they can to be healthy and fit.

Less than 10 percent of all heart attacks occur during exertion, which means that 90 percent do not involve exertion. Some heart attacks occur during sex or while on the toilet. Although many worry about the risks of exercise, no one ever suggests that we give up sex or trips to the toilet. The majority of cardiac fatalities occur in people over 45 years of age. Other risk factors include gender (male), obesity, elevated coronary risk factors (cholesterol, blood pressure), smoking, and physical inactivity. Physically inactive people are 56 times more likely to experience a problem during exertion. For this reason, the primary strategies for minimizing risk are (1) appropriate screening of prospective participants to identify those with a high risk of coronary artery disease and (2) control of exercise intensity, particularly during the first few weeks of an exercise program (Foster and Porcari 2001).

When patients take a stress test in a medical setting, the risk is one cardiac episode per 187,000 person-hours of testing. The average test lasts 15 minutes, so the odds of an event are 1 in 788,000, or about 1/100 of the chance of being in a major automobile accident per hour of driving. But when presumably healthy (asymptomatic) people are tested, the risks are much lower. The risk of death while jogging has been estimated at one death per year for every 396,000 person-hours of jogging. The average recreational runner averages three hours per week, resulting in a 1 in 2,600 chance of dying each year. Meanwhile, running reduces the risk of chronic disease 4 to 10 times (American College of Sports Medicine 2005). Clearly, the health benefits of physical activity outweigh the risks.

To make the transition from a sedentary lifestyle to an active one, we recommend six to eight weeks of walking most days of the week, slowly increasing distance and then pace until the person can walk 3 miles (4.8 km) in 45 minutes. When that prescription is filled, the risk of a heart attack will be minimized. Those who burn sufficient calories in physical activity cut their risk in half, when compared with those who are sedentary. Those who do some vigorous activity each week have 30 percent less chance of experiencing a heart attack than sedentary people. During vigorous exercise, the risk of a heart attack for the habitually active person rises slightly above the risk level of the resting sedentary person, but only during the period of exercise (Siscovick, LaPorte, and Newman 1985). For more on the risks and benefits of exercise see the Joint Position Statement coauthored by the American College of Sports Medicine (ACSM) and the American Heart Association (ACSM 2007).

Overuse Injuries

Other risks of fitness training range from minor musculoskeletal problems to overuse injuries. The risks are low for low-intensity and moderate-duration activities. Overuse injuries are more common when distance is increased rapidly, with high-intensity (e.g., interval) training, and with eccentric exercise (e.g., downhill running). With

adequate warm-ups and cool-downs, stretching, days off for rest, and a moderate rate of progression, fitness training is remarkably free of risk. Fitness instructors certified by the American College of Sports Medicine should assist employees in the transition to fitness.

Overtraining

Excessive training of athletes in the absence of adequate rest leads to overtraining, a syndrome that is characterized by a persistent decline in performance and a host of related symptoms (chronic fatigue, staleness, irritability, decreased interest, and weight loss). Mood changes include a decrease in the sense of vigor and increases in tension, anxiety, dejection, hostility, and confusion. Physiological correlates of overtraining include a rise in exercise and recovery heart rates, elevated cortisol, decreased testosterone, increased cytokines and white blood cells, and a loss of tissue protein (Sharkey and Gaskill 2006).

CASE STUDY
Overtraining and the Military

Does intense military training cause the symptoms of overtraining? In this French study, male soldiers were studied during three weeks of combat training (Gomez-Merino et al. 2003). Continuous heavy physical activity and sleep deprivation led to energy deficiency. Saliva samples were analyzed for secretory immunoglobulin A (sIgA), the first line of defense against respiratory infections. Plasma samples were analyzed for cortisol, testosterone, cytokines, and catecholamines. The prolonged military training led to immune impairment via a decrease in mucosal immunity. The training lowered testosterone, a classic sign of overtraining, reflecting a general decrease in anabolic steroid synthesis as a consequence of the physical and psychological strain.

Benefits of Fitness

Throughout this book we have listed many of the benefits of habitual physical activity and fitness. These include a reduced risk of heart disease, stroke, diabetes, obesity, and some cancers. Regular activity also lowers cholesterol levels and blood pressure and helps people lose weight. Activity and fitness reduce anxiety, depression, and stress. Fitness training improves job performance and reduces the risk of low back and other lost-time injuries. Over the years, improved fitness helps people maintain independence and delays the onset of frailty. Recent evidence also suggests that physical activity can reduce the likelihood of Alzheimer's and dementia as we age.

Some of these benefits can be calculated; others are incalculable. We can measure the value of reduced absenteeism and attrition, improved performance, and productivity, but how do we measure a heart attack that doesn't occur? The cost of a nonfatal heart attack can run into hundreds of thousands of dollars for surgery, hospitalization, rehabilitation, possible disability, or early retirement. We do some things because we have to and others because we should. Physically demanding occupations require job-related fitness (or physical maintenance) programs.

CASE STUDY
Firefighter Fatality

The latest findings in the ongoing study of American firefighter fatalities, reported in the *Morbidity and Mortality Weekly Report* (April 28, 2006), from the U.S. Centers for Disease Control and Prevention (CDC) (www.cdc.gov/niosh/fire), were that sudden cardiac death remains the leading cause of death in the line of duty. Half of all deaths for volunteer firefighters were cardiac related, and 39 percent of deaths for career firefighters were from heart attacks. As in past years, the deaths were attributed to stress and overexertion. According to Dr. Ridenour, epidemiologist with the CDC's National Institute for Occupational Safety and Health (NIOSH), "The leading cause of death is sudden cardiac death, and we can reduce these risks by having fitness programs and annual physicals. Fire departments should consider mandatory annual fitness exams for firefighters."

Summary

This chapter discussed the development of a job-related fitness program. We expanded the concept of aerobic fitness to include the first and second lactate thresholds and defined sustainable fitness as the workload one can maintain throughout the working day. We indicated the components of the strength–endurance continuum and provided prescriptions for each component. We emphasized the importance of core training for lifting and back health, and we determined job-related elements of body composition, providing advice for ways to increase lean body weight or lose excess fat.

After a discussion of program issues, we concluded with a consideration of the risks and benefits of physical activity and training. The risk of cardiovascular disease, the nation's number one killer, cannot be ignored. Physical inactivity is a major risk for heart disease. The data show that the substantial health benefits of physical activity clearly outweigh the risks. Training improves job performance and reduces the likelihood of lost-time injuries. So why do employees in physically demanding occupations avoid annual tests and fitness programs? We don't know. To review a physical fitness policy, see appendix F, page 211, and see standard operating procedures in appendix G, page 213.

PART V
Job-Related Issues

© iStockphoto.com/Tim McCaig

As long as men are free to ask what they must,
free to say what they think, free to think what they will,
freedom can never be lost, and science can never regress.

J. Robert Oppenheimer

The final part of the book deals with workplace issues related to fitness, safety, health, and performance. In chapter 13 we examine heat stress, cold, and altitude and what employees can do to perform safely and well in each environment. We consider the use of respiratory protection devices in chapter 14 and their impact on the worker. Chapter 15 provides lifting guidelines designed to reduce the risk of back injuries among most workers. A key point that permeates the chapters in this part of the book is that training is important for the maintenance of health, safety, fitness, and performance, a point that is generally ignored in the workplace, in regulatory agencies, and in the courts.

The final chapter of the book considers legal challenges and decisions. We look at the reasons for legal challenge, courtroom verdicts, and the unintended consequences of the decisions. We discuss how testimony and verdicts sometimes twist, ignore, or counter established knowledge in work physiology. We also argue that biased testimony and faulty decisions, even when made with good intentions, are not helpful to the workplace or the people the laws were crafted to protect. Chapter 16 concludes with a proposal for an alternative to litigation that uses a collective approach, avoids confrontation and biased testimony, and saves taxpayers' money.

Chapter 13

Environmental Impacts

© JOSH REYNOLDS

Environmental factors such as heat, humidity, cold, altitude, and even air quality can have profound effects on health and performance in the workplace. Failure to consider these effects can lead to serious problems, even death. On the other hand, workers can adjust or acclimatize to the environment, enabling them to perform well and comfortably under a wide range of conditions. In this chapter we consider the problems caused by extremes of temperature, humidity, cold, and altitude to see how fitness, acclimatization, and planning can minimize their effects. We discuss air quality and respiratory protection in chapter 14.

This chapter will help you do the following:

- Anticipate the effects of temperature on health and performance.
- Take the appropriate steps to prevent heat stress and cold exposure.
- Understand how fitness enhances workers' ability to acclimatize and perform in challenging environments.

Heat Stress and Heat Disorders

Hard work is often demanding, but it is even tougher in the heat, in high humidity, or under a glaring sun. Under these conditions, heat stress can occur. Heat stress can be defined as a rise in core body temperature beyond safe limits, fluid loss, an inability to produce sweat, and the resultant trapping of heat in the body. When hard work and protective clothing are combined with temperature, humidity, radiant heat, and lack of air movement, a worker could be facing a life-threatening rise in core temperature. Evaporation of sweat is the main line of defense against heat. As sweat evaporates, it cools the body. If the fluids lost in sweat are not replaced, the body's heat controls break down and body temperature climbs dangerously. When the body can't cope with this heat burden, the person experiences heat disorders including heat cramps, heat exhaustion, and life-threatening heatstroke.

Heat cramps are painful muscle cramps that strike workers who sweat profusely. They are less likely when fluid intake is adequate and the diet includes bananas, oranges, and a sprinkling of table salt with meals. Treatment involves electrolyte drinks (tomato juice, sport drinks) and stretching to relieve the cramp.

Heat exhaustion, characterized by weakness, clammy skin, headache, and nausea, is caused by inadequate fluid intake, electrolyte loss, or both. The fluid loss causes a drop in blood volume that severely limits work output. The electrolyte loss reduces the muscles' ability to function. Treatment includes rest in a cool place and electrolyte drinks.

Heatstroke results from failure of the body's heat controlling mechanism (see figure 13.1). It is characterized by hot (often dry) skin, very high body temperature (106 °F [41 °C]), confusion, slurred speech, possible convulsions, and loss of consciousness. Heatstroke is a medical emergency, requiring immediate treatment. The victim should be rapidly cooled by soaking the trunk with cold water and fanning to promote evaporation. The victim should then be transferred to a medical facility as quickly as possible.

© David Joel/Photographer's Choice/Getty Images

Evaporation of sweat is the main line of defense against heat stress because as sweat evaporates, it cools the body.

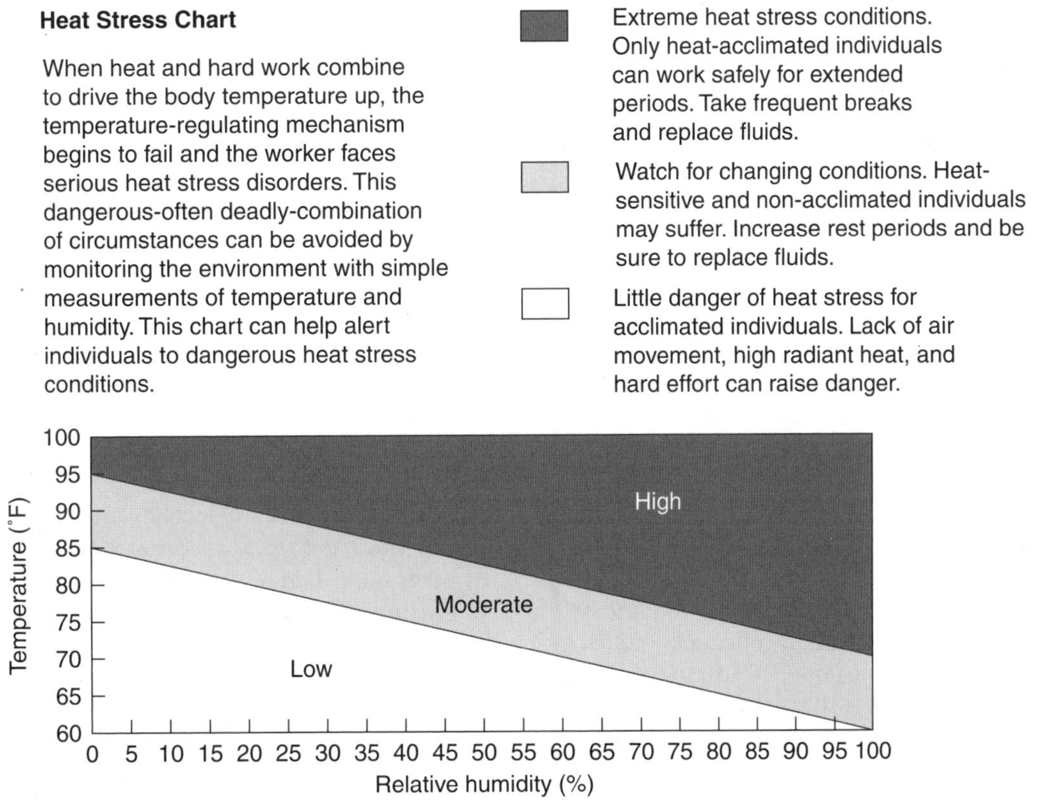
Heat Stress Chart
When heat and hard work combine to drive the body temperature up, the temperature-regulating mechanism begins to fail and the worker faces serious heat stress disorders. This dangerous-often deadly-combination of circumstances can be avoided by monitoring the environment with simple measurements of temperature and humidity. This chart can help alert individuals to dangerous heat stress conditions.
Extreme heat stress conditions. Only heat-acclimated individuals can work safely for extended periods. Take frequent breaks and replace fluids.
Watch for changing conditions. Heat-sensitive and non-acclimated individuals may suffer. Increase rest periods and be sure to replace fluids.
Little danger of heat stress for acclimated individuals. Lack of air movement, high radiant heat, and hard effort can raise danger.
High
Moderate
Low
Temperature (°F)
100
95
90
85
80
75
70
65
60
0 5 10 15 20 25 30 35 40 45 50 55 60 65 70 75 80 85 90 95 100
Relative humidity (%)

Figure 13.1 Heat stress chart.

From B.J. Sharkey, 1997, *Fitness and work capacity*, 2nd ed. (Missoula, MT: USDA Forest Service).

RHABDOMYOLYSIS

Exertional rhabdomyolysis is most frequently reported in military personnel (0.3 to 3.0 percent) and in law enforcement and firefighting trainees (0.2 percent) during hot, humid conditions. Damage occurs to muscle membranes, allowing cellular components (such as creatine kinase, myoglobin, and potassium) to leak out. Rhabdomyolysis can lead to renal failure, irregular heartbeats, and even death (Gaffin and Hubbard 2001). Risk factors include a sedentary lifestyle, high ambient temperatures, and intense exercise. Military and fire department records indicate that the risk goes down as fitness improves. People taking cholesterol-lowering statin drugs may be at higher risk, as are those with sickle cell anemia, those with some viral infections, and those susceptible to hyperthermia.

Preventing Heat Disorders

Although the recognition and treatment of heat disorders is important, the best approach is prevention. Employee actions in the months, weeks, and days before exposure are as important as the things they do when exposed to heat stress. A high level of aerobic fitness, with improved blood volume and circulatory efficiency, is the best protection against the heat. Fit workers start to sweat at lower body temperatures so they work with lower body temperatures. Fit workers also acclimatize twice as fast as unfit workers.

Acclimatization

Acclimatization is the process of adjusting to changes in the environment such as temperature and humidity. The body acclimatizes to the heat in 5 to 10 days, depending on the person's level of aerobic fitness. Once acclimatized, the body sweats

CASE STUDY

Uniformly Fit

The effects of variations in the wildland firefighter uniform were studied in the Human Performance Laboratory heat chamber (Cordes and Sharkey 1995). Subjects performed four randomly assigned laboratory trials on separate days, each at the same rate and grade on the treadmill and under the same environmental conditions (90 °F [32.2 °C], 50 percent relative humidity). The fittest worker ($\dot{V}O_2$max = 68 ml · kg^{-1} · min^{-1}) finished the two-hour work test in the heat chamber with a heart rate of 118 beats per minute, whereas the least fit ($\dot{V}O_2$max = 45 ml · kg^{-1} · min^{-1}) labored at 164 beats per minute. There was a highly significant inverse relationship between aerobic fitness and the working heart rate ($r = -.91$); the higher the fitness level, the lower the heart rate. Fitness differences overshadowed the effect of uniform variations (including no undershirt, short or long-sleeve undershirt, or a short-sleeve undershirt and a nomex facial shroud) on physiological responses during work.

at a lower temperature, increases sweat production, improves blood distribution, and operates at a lower heart rate. Workers can acclimatize by working one to two hours a day in the heat, taking breaks as needed, and replacing fluids. Acclimatization persists for several weeks, especially with regular physical training. Fitness and acclimatization are required to work effectively in hot or humid conditions and to perform in protective clothing.

To defend against heat stress on the job, crew bosses and workers should know when it is likely to strike. Several heat stress indexes use temperature and humidity to predict dangerous conditions. When heat stress conditions are elevated, workers should maintain hydration and modify their work behavior, if necessary.

Hydration

Because thirst is not an accurate indicator of fluid needs, workers should drink before they get thirsty. In hot weather, fluid replacement is essential before, during, and after work. It is common to lose more than a liter of sweat for each hour of work in the heat. In a hot, humid environment sweat rates can approach 3 liters an hour for short periods. Maximum sweat loss during a shift can be 8 to 12 liters. Adequate replacement of water, salt, and potassium is vital to maintain work capacity and to avoid heat disorders. With one liter of water loss, core temperature increases more than 0.5 °F, heart rate increases eight beats per minute, and cardiac output declines, making it difficult to work effectively.

For example, wildland firefighters generate over 400 kilocalories of heat each hour of work, and the environment and the fire contribute an additional 180 kilocalories of heat. To avoid heat storage, a firefighter must lose, on average, 580 kilocalories of heat each hour. Complete evaporation of 1 liter (1.057 quart) of sweat eliminates 580 kilocalories of heat. Adequate replacement of water, salt, and potassium is vital to maintain work capacity and to avoid heat disorders. That's why wildland firefighters and soldiers in the U.S. Army are encouraged

||||||||||||| MONITORING ||||||||||||| HYDRATION

Workers can assess their level of hydration by observing the frequency, volume, and color of their urine. If they stop urinating, they are dehydrated. Urine volume should be considerable if they are drinking before, during, and after work. If volume is low and it hurts to urinate, they are becoming seriously dehydrated. Urine color can also be used to assess hydration. Dr. Lawrence Armstrong correlated urine color to measures of hydration, including urine specific gravity and urine osmolality. The urine color was correlated to hydration status and to body water loss. When the urine is pale yellow, straw colored, or the color of wheat, workers are hydrated. If urine color is darker, they are becoming dehydrated (Armstrong 2000).

To prevent dehydration:

Before work: Drink one to two glasses of water or juice.

During work: Drink 1 cup (water or sport drink) every 15 minutes (1 liter per hour). Replace fluids during meal breaks.

After work: Drink juice or a sport drink after work and continue replacing fluids throughout the evening.

Adequate replacement of water, salt, and potassium is vital to maintaining work capacity and avoiding heat disorders.

to drink up to a liter of fluid (water or sport drink) for every hour of work in the heat. It's important to be aware that drinking larger quantities of water may contribute to low serum sodium levels or hyponatremia.

Hyponatremia

Excess water intake during prolonged exertion can disturb the body's fluid–electrolyte balance and lead to abnormally low levels of plasma sodium (less than 135 mmol/L) and a condition called hyponatremia. When excess water intake is combined with a loss of sodium in sweat, the risk of hyponatremia grows. A pronounced drop can cause unusual fatigue, disorientation, and severe headache; even death is possible if plasma sodium falls below 120 mmol/L. Hyponatremia occurs in prolonged distance events and in military training. For example, nine U.S. marines experienced hyponatremia after drinking 10 to 22 quarts of water over a few hours of exertion. All survived after emergency treatment (Gardner 2002). Workers can avoid hyponatremia by drinking up to 1 liter of fluid per hour, using a mix of electrolyte drinks (sport drinks) and water.

Sport Drinks

Urine production is reduced and fluid retention enhanced when fluids include some electrolytes (sodium, potassium), ensuring more blood volume for cardiac output and more water for sweat production. Studies of wildland firefighters indicate that the use of commercially available sport drinks containing carbohydrate and electrolytes helps maintain performance. For daylong work, we recommend that a quarter to a half of each liter consumed be a sport drink. Workers drink more when the drink is flavored. The carbohydrate in sport drinks maintains blood glucose, cognitive and immune function, mood, and work output. The electrolytes help retain the fluid (reduce urination) needed for sweat (Ruby et al. 2004).

PASS THE ELECTROLYTES

Workers in physically demanding occupations should use the salt shaker at mealtime for more sodium and eat bananas, oranges, and fruit juices for more potassium. Studies of athletes suggest the value of higher sodium levels in carbohydrate/electrolyte drinks during work in the heat. Workers should avoid salt tablets, but they can use electrolyte pills with water during prolonged hard work in the heat. They should not continue the elevated salt intake when they return to less arduous working conditions.

Studies do not suggest that one sport drink is superior to others, so long as the carbohydrate content is approximately 6 to 8 percent per liter. Speed of absorption is not important when working at or below 50 percent of $\dot{V}O_2$max. Energy bars and candy bars can provide additional sources of carbohydrate during work.

Pacing

People differ in their responses to heat stress, and some are at a greater risk for heat disorders than others. The reasons include low fitness, very high or very low sweat rates, excess salt loss in sweat, obesity, illness, previous heat disorders, some medications, and all recreational drugs, but especially methamphetamines. Because of these differences in heat tolerance, workers must pace themselves during work.

Work and rest cycles can be adjusted to prevent fatigue and heat stress. Shorter work periods and more frequent rest periods in a cool, shaded area or air-conditioned enclosure can minimize heat buildup. When possible, workers should perform the hardest work during the cooler morning or evening hours, change tools or tasks to minimize fatigue, and take frequent rest breaks.

If several days of hot work are required, employers should monitor workers' hydration and rehydration. Workers' body water loss can be measured by weighing them on a scale at the beginning and end of each working day. Weight loss should not exceed 1.5 percent of workers' total body weight in a workday. If it does, workers should increase their fluid intake. Dehydrated workers should not return to work.

Protective Clothing

Modern fire-resistant garments, designed to protect against sparks, embers, and direct exposure to flame, do so at a price in terms of heat stress. The fire-protective fabric reduces airflow and evapora-

CASE STUDY

Indexes for Heat Illness

Male and female U.S. marines were studied to determine whether continuous hot weather training resulted in cumulative heat stress (Wallace et al. 2005). Researchers used the WBGT (wet bulb globe temperature) heat stress index, which factors wet and dry bulb temperatures and the temperature of a black copper globe (WBGT = wet bulb × 0.7 + globe × 0.2 + dry bulb × 0.1). The risk of exertional heat illness increased with WBGT at the time of the event and with the previous day's WBGT. The results provide evidence for a cumulative effect of the previous day's heat exposure on heat illness.

Researchers at the U.S. Army Research Institute of Environmental Medicine developed a physiological strain index (PSI) to evaluate heat stress. The PSI is based on the heart rate and rectal temperature recorded during work. In the past, the rectal temperature was measured with a probe inserted in the rectum. Now it is possible to swallow a pill that transmits core temperature to a recorder worn on the belt. When used in hot-dry and hot-wet environments, the PSI differentiated between the two climates (Moran, Shitzer, and Pandolf 1998).

Marines dressed in level B (maximum respiratory and splash resistant) suits move a barrel of hazardous material. Clothing worn by first responders and structural firefighters can interfere with heat dissipation. This is another important reason for selecting fit and acclimated workers who are able to tolerate the physiological strain.

Photo by Cpl. Warren Peace, U.S. Marines

tive cooling, adding to the heat stress caused by work, the environment, and the fire. When possible, personal protective clothing should strike a balance between protection and worker comfort. In regard to wildland firefighting, Australian researchers said: "Clearly, the task of firefighters' clothing is not to keep heat out, but to let it out" (Budd et al. 1996).

Clothing worn by structural firefighters and first responders interferes with the dissipation of heat, making it even more important to select fit and acclimated workers who are able to tolerate the physiological strain. Unfit workers will be able to perform useful work for only a few minutes, and they are more likely than fit workers to suffer heat disorders.

CBRN Protection

Chemical/biologic/radiological/nuclear protective (CBRN) clothing is worn with an approved self-contained breathing apparatus (SCBA) or respirator, depending on the threat. The first responder is enveloped in an impermeable membrane to protect against chemical (e.g., ricin), biological (e.g., anthrax), or radiological/nuclear threats. The clothing creates a microenvironment that magnifies the effects of exertion, heat, and humidity.

A British researcher evaluated the physiological loads associated with the evacuation of an underground railway (subway). Two groups of subjects performed the simulation, one wearing normal personal protective equipment (PPE) and one wearing CBRN suits; both used extended-duration breathing apparatuses. The subjects were required to descend from street to platform level, drop down to the tracks, and travel 0.62 miles (1 km) to the simulated rescue scene. Once there, they were expected to rescue a 165-pound (75 kg) casualty from a stationary train. The rescue to street level was done with the escalator power both on and off.

When the power was on, all subjects wearing PPE succeeded, but with the power off, success fell to 50 percent. Subjects wearing CBRN clothing were forced to proceed at a slower rate, reaching the limiting core temperature (39.5 °C [103 °F]) or depleting their air supply before finishing the task. Temperature and relative humidity (RH) inside the garments were 22 °C (71.6 °F) and 45 percent RH for PPE and 26 °C (78.8 °F) and nearly 100 percent relative humidity for CBRN clothing. Subjects wearing CBRN clothing were unable to complete one extraction before they reached physiological limits for core temperature or air supply (Wilkinson 2005). This study indicates that first responders required to wear CBRN protection should be required to demonstrate their ability to complete their duties successfully in the microenvironment created by the protective clothing.

MEDICAL MANAGEMENT PLAN FOR FIREFIGHTER COMBAT CHALLENGE

The Firefighter Combat Challenge is a competitive event composed of critical firefighting tasks. The medical management protocols are used to minimize the risk of exertional problems. The plan, developed by Dr. Jack Harvey, medical director of the Firefighter Combat Challenge, is used at all Firefighter Combat Challenge competitions.

1.0 Precompetition Briefing

1.1 Discussion of nutrition for anaerobic events, including nutrition of the day of competition (low fat, high glucose). Superhydration prior to the event.

1.2 Discussion of format for postcompetition recovery. Immediate removal of SCBA and turnout gear. Walk to recovery area and start oral hydration with cold water only; no sport drinks initially. If recovery is not prompt, then start IV hydration. As soon as recovered, start walking and stretching. Upon completion of recovery, continued hydration with water and sport drinks.

1.3 Discussion of postexercise soreness and rhabdomyolysis.

1.4 Answer questions from audience.

1.5 Emphasis on temperature, humidity, and hydration if wet bulb is high as per ACSM guidelines.*

(continued)

(continued)

2.0 Medical Team

2.1 Team should consist of athletic trainers (1 or 2) and paramedics or EMTs with intravenous line training (3 to 6). Physician supervision with sports medicine experience recommended.

2.2 Equipment should include a covered area out of the sun with six chairs and four to six cots. Several containers of ice water for drinking and cups, with adequate resupply available. Towels and ice water for cooling of participants or hose with sprayer. Vital sign monitoring equipment and arrangement for hanging IV bags. IV supplies (Ringers lactate, 16-gallon [60.5 L] catheters and tubing). Estimate 20 IV lines and catheters and 40 bags of Ringers at the event site with resupply available for hot, humid days.

2.3 Ambulance with ALS capabilities on standby at site.

3.0 Postevent Care

3.1 Upon completion of event, staff will assist with ambulation of competitor to recovery area while helping remove gear and turnouts. Ice towel placed on competitor's neck.

3.2 Staff notes time of arrival at recovery station and condition of firefighter and encourages oral hydration and external cooling.

3.3 After five minutes, if competitor is recovering well, continue oral hydration and start competitor walking to assist recovery. Instruct competitor in stretching and provide water or sport drink upon dismissal from area.

3.4 If at five minutes competitor is not recovering, move to observation area and continue close observation. Vital signs taken and consideration of IVs for some prostrate competitors may be appropriate at this time. If competitor is vomiting, initiate IV.

3.5 If not recovered at 10 minutes, start IV and give 2 to 3 liters rapidly. If competitor recovers, go to 3.1 above; if no recovery, consider transport to emergency room.

*For additional information on heat stress, see the American College of Sports Medicine position stand on exertional heat illness during training and competition (Armstrong et al. 2007).

Cold Conditions

Because workers generate heat during work and clothing can be worn for protection, cold temperatures do not pose as large a threat as that of heat. However, exposure to low temperatures and wind (see figure 13.2) can lead to frostbite, hypothermia, and even death. The body cuts off blood flow to the extremities during cold exposure, leading to discomfort and loss of dexterity. The large muscular activity of work can maintain or restore blood flow. This activity takes considerable energy, so those exposed to cold weather must maintain a reserve of energy for use during prolonged effort and to meet unforeseen emergencies. Excessive fatigue is the first step on the road to hypothermia.

Wind chill describes the effect of wind speed on heat loss. Cold temperatures or wind chill can cause frostbite, which is damage to the skin and underlying tissue. Workers need to protect sensitive areas to avoid frostbite and the pain that comes with rewarming. Hypothermia begins when the body loses heat faster than it can be produced. Fatigue and energy depletion compound the problem, as does rapid cooling from evaporation of sweat, snow, or rain. When the cold reaches the brain, the body begins to shut down. This is a medical emergency, and the victim should be transported to a medical facility as quickly as possible. Hypothermia often occurs at temperatures above 30 °F (–1 °C).

To avoid cold weather problems, workers should do the following:

- Dress in layers with wicking garments, insulation, and a weatherproof outer layer.
- Take layers off as they heat up.
- Wear a hat that protects the ears.
- Maintain energy level and avoid exhaustion.
- Acclimatize to the cold to minimize discomfort.

Although cold air will not freeze the tissues of the lungs, even at subzero temperatures, it may make strenuous work difficult for those prone to

Wind chill chart

Temperature (°F)

Wind (mph) / Calm	40	35	30	25	20	15	10	5	0	-5	-10	-15	-20	-25	-30	-35	-40	-45
5	36	31	25	19	13	7	1	-5	-11	-16	-22	-28	-34	-40	-46	-52	-57	-63
10	34	27	21	15	9	3	-4	-10	-16	-22	-28	-35	-41	-47	-53	-59	-66	-72
15	32	25	19	13	6	0	-7	-13	-19	-26	-32	-39	-45	-51	-58	-64	-71	-77
20	30	24	17	11	4	-2	-9	-15	-22	-29	-35	-42	-48	-55	-61	-68	-74	-81
25	29	23	16	9	3	-4	-11	-17	-24	-31	-37	-44	-51	-58	-64	-71	-78	-84
30	28	22	15	8	1	-5	-12	-19	-26	-33	-39	-46	-53	-60	-67	-73	-80	-87
35	28	21	14	7	0	-7	-14	-21	-27	-34	-41	-48	-55	-62	-69	-76	-82	-89
40	27	20	13	6	-1	-8	-15	-22	-29	-36	-43	-50	-57	-64	-71	-78	-84	-91
45	26	19	12	5	-2	-9	-16	-23	-30	-37	-44	-51	-58	-65	-72	-79	-86	-93
50	26	19	12	4	-3	-10	-17	-24	-31	-38	-45	-52	-60	-67	-74	-81	-88	-95
55	25	18	11	4	-3	-11	-18	-25	-32	-39	-46	-54	-61	-68	-75	-82	-89	-97
60	25	17	10	3	-4	-11	-19	-26	-33	-40	-48	-55	-62	-69	-76	-84	-91	-98

Frostbite times: 30 minutes, 10 minutes, 5 minutes

Figure 13.2 Wind chill chart.

Courtesy of the National Oceanic and Atmospheric Administration. www.noaa.gov

© Terje Rakke/The Image Bank/Getty Images

Workers need to protect sensitive areas to avoid frostbite and the pain that comes with rewarming.

airway constriction. Workers should use a mask or a scarf to minimize the effect of cold air on airways. For additional information, see the American College of Sports Medicine position stand on the prevention of cold injuries during exercise (Castellani et al. 2006)

Altitude Acclimatization

Although millions of people live at elevations above 10,000 feet (3,050 m), no permanent habitations are found above 18,000 feet (5,488 m), suggesting that such an elevation may be incompatible with adaptation and long-term survival. Indeed, the upper reaches of 29,000 feet (8,841 m) on Mount Everest have been called the death zone. Climbers recognize the need to train and acclimatize to succeed at altitude, and so should workers involved in physically demanding occupations.

As one ascends in elevation, atmospheric oxygen levels decline. Working at elevations below 5,000 feet (about 1,500 m) has little noticeable effect on healthy workers. As they ascend above that elevation, the reduced level of oxygen reduces their work capacity. Although highly fit workers can do more at altitude, they too are affected by the diminished oxygen supply.

Because a reduced oxygen supply affects their ability to take in, transport, and use oxygen,

Reduced oxygen supply at high altitudes affects a worker's ability to take in, transport, and use oxygen. Although it takes several weeks to make a good adjustment to altitude, a worker can improve with a week of altitude exposure for each 1,000 feet (300 m) above 5,000 feet (1,500 m).

workers should make a partial adjustment to the effects of altitude. Working at altitude leads to acclimatization by doing the following:

- Increasing air intake (ventilation)
- Improving oxygen transport (more red blood cells)
- Improving the use of oxygen in muscles (and increased capillaries)

These adjustments reduce—but never eliminate—the effects of altitude on aerobic fitness. Although it takes several weeks to make a good adjustment to altitude, one can improve with a week of altitude exposure for each 1,000 feet (300 m) above 5,000 feet (1,500 m). Athletes have learned that the best way to adjust to altitude is to "live high and train low"; that is, live at high altitude to maximize the effect of altitude on the production of red blood cells and train at low altitude to maintain intensity in the training program (Chapman, Stray-Gunderson, and Levine 1998). Workers may acclimate more rapidly if they live at the elevation of the job.

If workers must work above 5,000 feet (1,500 m), they should take it easy for the first few days, take frequent breaks, and avoid excessive fatigue.

They should eat a high-carbohydrate diet for energy and take care to maintain hydration because altitude hastens fluid loss. As with heat stress, people differ in their altitude tolerance. Above 8,000 feet (2,400 m) a few may begin to experience mild symptoms of acute mountain sickness, characterized by fatigue, headache, and lack of appetite. They may require more time to adjust.

Summary

Workers must be taught to anticipate the effects of the environment on health and performance and to take appropriate steps to minimize those effects. Sometimes the most effective steps must be taken weeks or even months before exposure. An elevated level of aerobic fitness enhances workers' ability to perform in hot conditions. Fit workers acclimatize more rapidly and retain acclimatization for longer periods of time than unfit workers. Fitness, acclimatization, and hydration are the ways to avoid heat stress while performing in high temperatures. First responders required to wear CBRN protection should be required to demonstrate the ability to work in the microenvironment created by the protective clothing.

Because workers generate heat during work and clothing can be worn for protection, cold temperatures do not pose as large a threat as that of heat and humidity. Exposure to low temperatures and wind can lead to frostbite, hypothermia, and even death, which is why workers need to protect sensitive areas to avoid frostbite.

Workers should adjust to the effects of altitude because reduced oxygen supply affects their ability to take in, transport, and use oxygen. Working at altitude leads to acclimatization via increased air intake, improved oxygen transport, and improved use of oxygen in muscles.

Chapter 14

Respiratory Protection

© AP Photo/Daily Press and Argus, Gillis Benedict

Respiratory protection is required in some physically demanding occupations such as structural firefighting and for first responders to chemical, biological, and radiological/nuclear emergencies. U.S. organizations that require respiratory protection must establish a respiratory protection program required by the Occupational Safety and Health Administration, or OSHA (29 CFR 1910.134), the agency that monitors workplace compliance with regulations. The program makes the following stipulation:

Persons should not be assigned to tasks requiring use of respirators unless it has been determined they are physically able to perform the work and use the equipment.

The OSHA program does not specify the test or examination that will establish the ability to work while wearing a respirator. This chapter explores the impact of respiratory protection on workers and suggests ways to establish and maintain workers' ability to work while wearing respiratory protection.

This chapter will help you do the following:

- Understand the health hazards of respiratory exposure.
- Review the elements of the respiratory protection program.
- Select the appropriate respirator for the exposure.
- Use pulmonary function, maximal voluntary ventilation, and aerobic fitness tests to establish worker capability.
- Compare the value of laboratory and field performance tests.
- Evaluate the impact of air-purifying respirators on women.

Respiratory Hazards

Work environments contain a number of health hazards ranging from irritating wood smoke to the life-threatening products of structural combustion. Some products, such as silica, coal dust, and asbestos, cause disability or death years after the exposure. First responders may be faced with chemical, biological, and radiological/nuclear exposures. Respirators are required in a number of occupations.

The biological effects of respiratory hazards include the following:

- Deadening of ciliary action, which normally clears particles out of the airway
- Suppression of the immune system
- Reduced oxygen transport (carbon monoxide)
- Chronic obstructive pulmonary disease
  - Irritation of airways in chronic bronchitis
  - Alveolar breakdown in emphysema
  - Loss of diffusing surfaces in pulmonary fibrosis
- Lung and other cancers
- Heart disease

Respirators are required in a number of occupations that require combating health hazards ranging from irritating wood smoke to the life-threatening products of structural combustion.

This is just a partial list, but you get the idea; there are many adverse effects of occupational pollution. Though some forms are troublesome, debilitating, or even fatal, no single source of pollution is as deadly as the cigarette, which causes all but one of the biological effects just listed (pulmonary fibrosis). Cigarette smoking is the largest preventable cause of death in the United States, responsible for over 400,000 deaths annually. The U.S. surgeon general's report, *The Health Consequences of Smoking: Cancer and Chronic Lung Disease in the Workplace* (1985), states that asbestos exposure increases the risk of lung cancer 5 times, smoking increases the risk 10 times, and smoking and asbestos exposure increase the risk 87 times (www.surgeongeneral.gov/library/reports.htm). Prohibiting smoking saves lives. Responsible employers ban smoking, on the job or off, especially if lung and heart disease are considered job-related illnesses. Why should they spend thousands of dollars on respiratory protection for workers who smoke, or pay for self-imposed lung or heart-related disabilities?

NO SMOKING

Fire organizations are concerned about the health and safety of their employees and spend many thousands of dollars for medical exams, lung function tests, and respiratory protection. They cannot determine the contribution of workplace smoke exposure to a respiratory problem, however, when the employee is a smoker. The U.S. Tenth Circuit Court of Appeals upheld a fire department's right to prohibit personnel from smoking, on and off the job. The suit was filed by a firefighter who was fired by the Oklahoma City Fire Department for smoking on an unpaid lunch break that was not taken on department property. The court ruled that an employee's freedom can be infringed on if there is a rational relationship between a rule and a legitimate concern of the department.

Respirator Selection

Respiratory hazards are complex. The smoke from forest fires, for example, contains hundreds of products. Structural firefighters are exposed to thousands of hazards in smoke, many of them toxic, in a confined environment. In some cases workers, such as those working at ground zero after September 11, 2001, are unaware of the hazards present in their work environment, making them more vulnerable to injury. In the United States the National Institute for Occupational Safety and Health (NIOSH) conducts research and makes recommendations for the prevention of work-related illness and injuries. NIOSH recommends that air sampling be conducted to determine exposure levels found in the workplace. Once the hazards have been identified, appropriate respiratory protection should be selected.

After all hazards have been identified, the criteria for respirator selection have been evaluated, and the requirements and restrictions of the respiratory protection program have been met, employers should use the NIOSH respiratory selection logic sequence to identify the class of respirators that should provide adequate respiratory protection (www.cdc.gov/niosh). If the respirator is intended for use during firefighting, a full-facepiece self-contained breathing apparatus (SCBA) that meets the requirements of the National Fire Protection Association (NFPA) must be employed. An exception to this requirement is wildland firefighting, in which the smoke has not been found to be immediately dangerous to life and health (IDLH) (Reinhardt and Ottmar 1997).

An approved SCBA should be used in an oxygen-deficient atmosphere (less than 19.5 percent). The respiratory protection logic sequence continues until the type of respirator, or escape-only device, has been determined. Air-purifying respirators use filters to remove particulate and absorbents to remove vapors. In all cases in which the employer requires workers to wear respirators, even if their exposure is below the occupational exposure limit, OSHA (2002) requires that the employer establish and implement a written respirator protection program.

Respiratory Protection Program

The OSHA program requires written operating guidelines covering the selection and use of respirators for each task or operation for which they are employed. The respiratory protection program should include the following:

- Respirator selection based on the hazards
- Training, fit-testing, and instructions in the use and limitations of respirators
- NIOSH-approved respirators, when available
- Respirator assignment to individuals, when practical
- Cleaning instructions and schedule
- Storage in a convenient, clean, and sanitary location
- Inspection and maintenance procedures and schedule
- Monitoring of work area conditions and the degree of employee exposure
- Inspection and evaluation to assess effectiveness
- Medical evaluation to determine workers' ability to perform work and use the equipment (respirator)

Employees should not be assigned to perform tasks that require the use of respirators unless

CASE STUDY
Exposure to Smoke

It was early August, and the growing haze announced the emergence of forest fires in the vicinity of a western U.S. city. Within a few days the haze darkened to an acrid cloud that burned the eyes and obscured the view of the local hills. Particulate matter (PM10) more than tripled the U.S. Environmental Protection Agency's (EPA) 24-hour PM10 standard of 150 $\mu m/m^3$, and the fires were growing. The media carried public health announcements cautioning residents to stay inside and avoid the health hazards of smoke. Outdoor workers at the local university heard the health warnings and decided that they shouldn't be working in such a dangerous atmosphere. When the university environmental health officer reviewed the Occupational Safety and Health Administration (OSHA) exposure limits that apply to the workplace, he found that the particulate and carbon monoxide exposure levels were about one-tenth of the permissible exposure limits. Reluctantly, the crew returned to work, some wearing dust masks.

they have been determined to be physically able to perform the work while using the equipment. A physician should determine what health and physical conditions are pertinent. The respirator user's physical condition must be reviewed periodically. OSHA does not designate the way to assess an employees' physical ability to perform work and use the respiratory protective device. We provide several options in the following section.

Medical Evaluation

At present no test or battery of tests can unequivocally determine the ability to work with an air-purifying respirator. Various approaches have been used, including pulmonary function and the maximal voluntary ventilation test. Other options include a test of maximal oxygen intake ($\dot{V}O_2$max) and a job-related field test.

Pulmonary Function

In a pulmonary function test, the subject takes a maximal inhalation and then exhales—as fast as possible—through the valve of the analyzer. The device calculates a number of measurements, including the following:

- Forced vital capacity, maximal amount of air exhaled (FVC in liters)

- Forced expiratory volume in one second (FEV1)

- FEV1-to-FVC ratio (FEV1/FVC)

- Forced expiratory flow in midexhale (FEF 25-75)

- Peak expiratory flow rate in liters per second (PEFR)

The test can also involve inspiratory measurements (e.g., peak inspiratory flow rate or PIFR). A pulmonary deficit or airway obstruction is defined as when a measure is less than 70 percent of expectations (based on age and height) (see figure 14.1). Surprisingly, lung function measures predict little about athletic or work performance.

In one study, subjects were evaluated for their ability to perform prolonged work while wearing a respirator (Sharkey and Mead 1993). The results were correlated to measures of pulmonary function. The respirator was a half-face device fitted with a high-efficiency particulate air filter (HEPA)

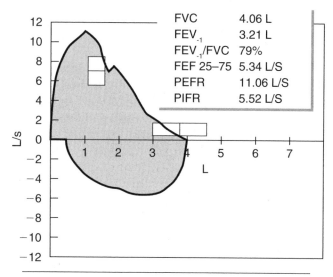

Figure 14.1 Lung function test results.

From B.J. Sharkey, 1997, *Health hazards of smoke: Recommendations of the Consensus Conference* (Missoula, MT: USDA Forest Service).

and organic vapor/acid gas absorbent (OV/AG). PIFR was also measured to mimic the effort involved in pulling in air against the resistance of the respirator canister. The lung function measurements were correlated to work performance (r = .6 to .73). The PIFR, while significantly related to performance, was not a better predictor than other pulmonary function measurements (r = .68).

Maximal Voluntary Ventilation

The MVV test requires rapid and deep breathing for 12 or 15 seconds. The results are then extrapolated to the volume that would have been breathed had the subject continued for one minute. The MVV is usually 25 to 33 percent higher than maximal ventilation during a $\dot{V}O_2$max test (e.g., max ventilation on $\dot{V}O_2$max test = 120 liters of air per minute; MVV = 180 L/min). In this respect, the ventilatory capacity is overbuilt for the demands we place on it during work or sport.

In a study by Wilson and Raven (1989), the MVV test was correlated to the performance of prolonged work while wearing a respirator. The respirator significantly reduces air intake, so the MVV value was first adjusted for the effects of the respirator:

$$MVV_{adj} = (MVV \times 0.49) + 29$$
$$[\text{e.g.,} \ 180 \times 0.49 = 88.2 + 29 = 117.2]$$

The adjusted MVV was then reduced by one-half to reflect the effect on daylong work output. If the final score fell below the ventilatory cost of wildland firefighting (40 to 60 L/min), the candidate would be predicted to have difficulty working all day with the respirator.

$$MVV_{adj} \times 0.5 = \text{daylong value}$$
$$[\text{e.g., } 117.2 \times 0.5 = 58.6 \text{ L/min}]$$

The study confirmed the value of the test as a measure of lung function while using a respirator for prolonged work, but the correlation with performance was not superior enough to other tests to warrant the use of the procedure for the selection of workers ($r = .60$). However, we can conclude that subjects with MVV values below 100 L/min would be severely taxed if forced to perform all day while wearing a respiratory protective device (adjusted MVV falls below 40 L/min) (Sharkey 1997c).

Aerobic Fitness

Aerobic fitness is defined as the ability to take in, transport, and use oxygen, so pulmonary function is an important part of the measurement. Measured $\dot{V}O_2$max and a prediction of aerobic fitness have been significantly correlated to performance with respirators and to measures of pulmonary function.

The American Industrial Hygiene Association (AIHA) recommends that a respirator be worn for at least 30 minutes, and during part of this time workers should exert themselves to the level that would be required on the job (AIHA 2006).

The wildland firefighter pack test (3-mile [4.8 km] hike with 45-pound [20.4 kg] pack in 45 minutes) predicts aerobic fitness. The energy cost of the test is equivalent to the energy cost of the job. Performance of prolonged work while wearing a respirator was significantly correlated to performance on the pack test ($r = .71$). Males and females who completed the pack test in 45 minutes or less had sufficient pulmonary capacity and were not adversely affected by the air-purifying respirator (Sharkey 1997c).

Work Performance

Air-purifying respirators (APRs) have been shown to decrease work performance as a result of breathing resistance, increased dead space, heat stress,

and respirator weight. They increase the sense of breathlessness (dyspnea) during strenuous effort and have been shown to cause claustrophobia in some subjects. Breathing resistance and reductions in work performance were directly related to respirator resistance (Thompson and Sharkey 1966). More recent studies compared the effect of modern APRs on work performance.

Arm work (cranking) was not reduced significantly with a respirator. This outcome was surprising because studies have shown diminished levels of pulmonary ventilation during work with the arms. The results did show a significant reduction in arm peak $\dot{V}O_2$ and peak ventilation. The decline in work performance while wearing the respirator was 4 percent for men and 8.3 percent for women. Subjects in this study used an upper-body exercise device (arm cranking) to isolate the arms and allow an accurate measurement of work performance. Work with hand tools often involves the arms, trunk, and legs, often with trunk flexion that could restrict pulmonary ventilation. An additional study using fire line construction tools found that upper-body work was not significantly reduced while wearing a respirator (Rothwell and Sharkey 1996).

APRs decreased treadmill work performance significantly in one hour (16.3 percent), reduced both maximal and prolonged work performance, and blunted the pulmonary response to vigorous work on the treadmill (see figure 14.2). When identical masks equipped with different cartridges (HEPA versus HEPA + OV/AG) were compared, the decline in performance with the respirator was proportional to the breathing resistance. (It should be noted that, in general, resistance increases with respiratory protection.) The HEPA filter protects against the inhalation of particulates. The addition of OV/AG absorption doubles the breathing resistance and doubles the decline in work performance (Sharkey 1997c).

Additional protection against carbon monoxide could be achieved, but at a considerable physiological cost. Converting carbon monoxide to carbon dioxide is an exothermic reaction that raises the temperature of the inspired air, increasing the breathing rate and sense of fatigue. An increase in carbon dioxide, the main respiratory stimulus, further increases pulmonary ventilation. Finally, the material used to remove carbon monoxide adds to the resistance of the device, causing an even greater decline in performance.

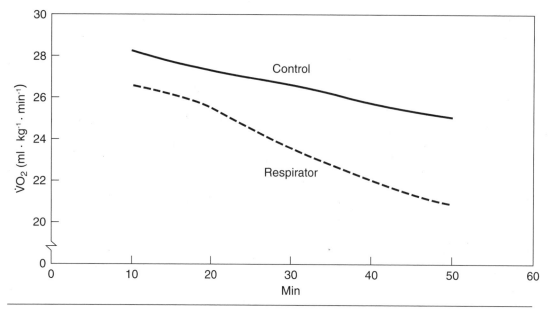

Figure 14.2 Effect of respiratory protection on work performance.

Adapted from B.J. Sharkey, 1997, Respiratory protection. In *Health hazards of smoke: Recommendations of the consensus conference*, edited by B.J. Sharkey (Missoula, MT: USDA Forest Service), 65-74.

CASE STUDY

Hazards in Smoke

Breathing zone studies of wildland firefighters demonstrate relatively low exposure to hazards. Respirable particulate was 13.8 percent of the permissible exposure limit, and carbon monoxide was 8.2 percent of the limit. In the numerous wild and prescribed fires studied, a small number of cases exceeded OSHA permissible exposure limits (see table 14.1).

A risk assessment determined little health or cancer risks at average levels of exposure, and that health effects were moderate and reversible. The exposures are intermittent and seasonal, allowing the lungs time to recover. Healthy lungs have a remarkable ability to recover from short-term exposure. Based on the exposure data, health effects, and the risk assessment, participants at a consensus conference voted to use a risk management approach to limit exposure, but not to recommend the adoption of respiratory protection for wildland firefighters. Respirators remove particulate and gases, but they do not remove carbon monoxide. Subjects in respirator field trials complained of headaches from excess carbon monoxide inhalation. Recommendations for risk management included improved training and tactics, monitoring, health maintenance, and medical surveillance (Sharkey 1997c).

Table 14.1 Wildfire and Prescribed Fire Breathing Zone Samples

Source	Wildfire	Prescribed fire	OSHA*	NIOSH
Particulate	0.69 µg/m^3	0.63	5.0	—
Carbon monoxide	4.1 ppm	4.1	50	35**
Formaldehyde	0.023 ppm	0.047	0.75	0.016
Acrolein	0.003 ppm	0.009	0.1	0.1
Benzene	0.016 ppm	0.016	1.0	0.1

*OSHA eight-hour permissible exposure limits (PEL): A healthy worker should be able to work within that standard without significant health effects.

**The OSHA standard was revised to 35 ppm in the 1980s until a court challenge forced the agency to withdraw the more stringent standard.

Adapted from T. Reinhardt and R. Ottmar, 1997, Employee exposure review. In *Health hazards of smoke: Recommendations of the Consensus Conference,* edited by B.J. Sharkey (Missoula, MT: USDA Forest Service), 29-40.

Senior airman Jessica Mueller, 460th Medical Group Bioenvironmental Flight, programs the high-volume air sampler, which samples airborne alpha radiation at the center of a contamination point. Because women's pulmonary function capacities are typically lower than those of men, it is important to understand the effect of air-purifying respirators on their ability to perform arduous work.

Protection should be appropriate to the exposure. For some subjects, a mask with no cartridges (and no resistance) can lower performance as a result of psychological factors. During hard work with a respirator, some subjects become claustrophobic and consequently refuse to work with the device.

APRs and Women

An extensive review of the literature revealed few studies in which women had been included as subjects in respirator studies. Because women comprise a significant percentage of the firefighter workforce, and because their pulmonary function capacities are, on average, lower than those of men, it is important to understand the effect of air-purifying respirators on their ability to perform arduous work.

Pulmonary function measures are associated with body size. The average value for women's forced vital capacity is 67 percent of men's FVC value (3.7 versus 5.5 L), and for maximal voluntary ventilation is 72 percent of men's MVV value (131 versus 182 L/min). Men and women were studied in arm and leg work while wearing respirators. Women averaged 53.1 percent of men on the arm ergometer test, reflecting the difference in upper-

body strength (50 to 60 percent of men's values). On treadmill tests the women scored 87.9 percent of men's $\dot{V}O_2$max values (43.4 versus 49.4 ml · kg^{-1} · min^{-1}), a difference that was not statistically significant.

Women who passed the pack test in 45 minutes ($\dot{V}O_2$max of 45 ml · kg^{-1} · min^{-1}) demonstrated sufficient pulmonary capacity and were not adversely affected by wearing APRs (Sharkey 1997c). However, it is clear that women and men with low values for $\dot{V}O_2$max, lung function, peak inspiratory flow rate, or maximal voluntary ventilation may indeed have difficulty performing prolonged work while wearing respirators. Unfortunately, few organizations that use respiratory protection apply performance standards for their workers. The use of respiratory protection provides yet another reason for requiring career-long maintenance of aerobic fitness and work capacity.

Summary

This chapter included a case study about exposure to forest fire smoke. The community exposure levels, sometime averaging above 300 µg/m^3, far exceeded the U.S. Environmental Protection Agency's (EPA) 24-hour standard for particulate matters (150 µg/m^3). However, the community exposure was less than half the average exposure for wildland firefighters, and the firefighters were well below the OSHA eight-hour particulate standard (5,000 μg/m³). The EPA standard is designed to protect those who are most susceptible, the very young, the old, and those with preexisting heart or lung disease. The OSHA standard focuses on the healthy worker, who should be able to work up to the exposure limit without adverse health effects. The OSHA standard is an eight-hour standard, but even if we divided the standard by 3, it would still be 11 times higher than the EPA 24-hour standard (5,000 / 3 = 1,667 versus 150 μg/m³).

This chapter outlined some of the biological effects of toxins to help explain the health hazards of respiratory exposure. We noted how cigarette smoking is the largest preventable cause of death in the United States. Smoking can exacerbate the effects of other toxins. According to a U.S. surgeon general's report, asbestos exposure increases the risk of lung cancer 5 times, smoking increases the risk 10 times, and smoking and asbestos exposure increase the risk 87 times.

We reviewed some of the regulations governing the use of respirators, including the need for a medical evaluation. Pulmonary function tests help to establish a worker's ability to work with a respirator, as does the maximal voluntary ventilation test. But so does a laboratory test of $\dot{V}O_2$max or a field performance test. We recommend that the medical evaluation include a test that reflects the physical demands of the job.

Air-purifying respirators decrease work performance as a result of breathing resistance, increased dead space, heat stress, respirator weight, and psychological factors. For this reason, workers need sufficient physical capacity to work with the device. This is particularly true for women, whose pulmonary function and maximal voluntary ventilation values are lower than men's. Women with a $\dot{V}O_2$max of 45 ml $\cdot$ kg^{-1} $\cdot$ min^{-1} have sufficient capacity to perform physically demanding work while wearing respirators.

Chapter 15

Lifting Guidelines

© Kelvin Murray/Stone/Getty Images

Common occupational tasks such as lifting, carrying, twisting, turning, pushing, and pulling often lead to musculoskeletal injuries. Injuries to the back account for about 20 percent of all injuries and illnesses in the workplace and approximately 25 percent of workers' compensation costs. Overexertion is blamed for most occupational injuries, accounting for over 30 percent of all injuries.

This chapter considers the establishment of safe lifting guidelines for workers. Researchers establish safe lifting loads based on psychophysical, biomechanical, and physiological standards. We review these standards and examine the most common manual material handling assessment tool, the NIOSH lifting guidelines. In the process we identify the risks of lifting and the strengths and limitations of the guidelines and provide recommendations to improve safety and performance in physically demanding occupations.

This chapter will help you do the following:

- Understand how lifting guidelines are established and discuss several lifting standards.

- Understand how the revised NIOSH lifting equation works and identify its limitations.

- Assess the impact of lifting guidelines on selection and training procedures.

Lifting Standards

We are all aware of common lifting guidelines. Workers are told not to carry more than one-third of their body weight in a backpack, a standard difficult to apply in the military, where packs often exceed 70 pounds (31.8 kg). The military and mountaineers carry more than 70 pounds, and backpackers often carry more than one-third of their weight all day long. Where do lifting guidelines come from? Lifting standards are based on psychophysical, biomechanical, and physiological studies.

Psychophysical Standards

Industry researchers have attempted to design material handling tasks suited to the capabilities of the industrial population. The psychophysical approach involves asking a sample of workers to determine the maximum acceptable weight (MAW) for a particular task, such as lifting a box from the floor to the level of a truck bed. Typically, the evaluation seeks a load that is acceptable to the majority (95 percent) of workers. Using loads selected by workers results in substantially reduced injury rates.

Biomechanical Standards

Biomechanical models estimate the strain of compressive forces on the spine during work tasks. Low back compressive forces have been estimated for a number of lifting positions, distances, and frequencies. The major contributor to spine compression forces is the weight of the load. Compression forces increase as the load is moved away from the body or to the side of the body. Low back pain is related to disc compression forces estimated by biomechanical models. The relevant factors have been included in the NIOSH lifting guidelines.

Physiological Standards

The physiological approach focuses on muscular strength and endurance and the energy demands of the task. Strength and muscular endurance are essential to avoid overexertion and injury. Low aerobic fitness may lead to fatigue and injury. For

U.S. Marine Corps photo by Cpl. Peter R. Miller

Military personnel and mountaineers carry more than 70 pounds (31.8 kg), which is more than one-third of their weight, for much of the working day.

repeated lifting, as in moving loads or working with hand tools, energy and muscular requirements are combined to set the limits of work capacity. Workers with high levels of aerobic fitness can work at higher work rates than those with lower levels of fitness; that is, they can do more contractions per minute. For stronger workers the load is a lower percentage of their maximal strength. They can lift more per repetition or do more contractions with lesser loads (see figure 1.2, page 7).

For repeated lifts, the load should not exceed 20 percent of the person's muscular strength. In many physically demanding occupations, the load cannot be redesigned; at times it can't even be anticipated. A firefighter may be trained to rescue the average adult, but rising levels of overweight and obesity could result in the firefighter being confronted with a 300-pound (136 kg) victim.

Regardless of how lifting standards are developed, one thing is perfectly clear: Those with higher levels of muscular and aerobic fitness are better able to do the job and far less likely to be injured than those with lower levels. Low back pain sufferers have less muscular strength and endurance than nonafflicted workers. Indeed, loads considered hazardous to the average worker in biomechanical models are easily sustained by physically fit workers.

© Johner/Johner Images/Getty Images

For repeated lifting, as in moving loads or working with hand tools, energy and muscular requirements are combined to set the limits of work capacity. Workers with high levels of aerobic fitness can work at higher work rates than those with lower levels of fitness.

IIIIIIIIII THE SHRINKING IIIIIIIIII SPINE

A reduction in stature has been used as a measure of the load on the spine. Musculoskeletal influences of aging may influence a person's response to compressive loads on the spine and the resulting loss of stature. Precision stadiometry was used to assess spinal shrinkage in young and older subjects following a regime of weightlifting (Reilly and Freeman 2006). Each group performed two sets of 12 exercises, with loads established relative to each person's capability. Both groups showed a similar pattern of spinal shrinkage. The authors concluded that healthy older workers are not compromised by their age in activities that include handling and lifting of weights, so long as the loading is related to individual capability. The challenge is to develop and maintain each worker's capability.

NIOSH Lifting Equation

The U.S. National Institute for Occupational Safety and Health (NIOSH) lifting equation was designed to assist in the development of ergonomic solutions designed to reduce the physical stresses associated with manual lifting. First introduced in 1981, the revised lifting equation reflects new findings and provides methods for evaluating asymmetrical lifting tasks and lifts of objects with poor coupling between the worker's hands and the object. Although the revised lifting equation has not been fully validated, the recommended weight limits are consistent with or lower than those generally reported. The authors claim that the revised equation is more likely to protect healthy workers in a wide variety of lifting tasks. They stress that the equation is but one tool of many used to prevent work-related low back pain and disability (Waters et al. 1993).

Limitations of the Equation

The NIOSH lifting equation is a tool for assessing the stress of two-handed lifts. It does not account for nonlifting tasks such as holding, pushing, pulling, carrying, walking, and climbing. If nonlifting activities account for more than 10 percent of total activity, measures of energy expenditure or heart rate may be required to assess the demands of the tasks. The equation does not account for unexpectedly heavy loads, slips, or falls. The equation assumes that the foot/floor surface coupling provides sufficient friction to avoid slipping. If the environmental conditions are unfavorable (temperature below 66 or above 79 °F [18.9 to 26.1 °C]; humidity below 35 or above 50 percent), other measurements may be necessary (core temperature, heart rate, energy expenditure) to gauge the effects of the environment.

The NIOSH equation does not apply if lifting and lowering occur in any of the following ways:

With one hand

For over eight hours

While seated or kneeling

In a restricted work space

With unstable objects

While carrying, pushing, pulling

With wheelbarrows or shovels

With high-speed motion (>30 in./sec [76 cm/sec])

With unreasonable foot/floor coupling

In an unfavorable environment

Based on these limitations, the equation is not appropriate for most physically demanding occupations such as structural and wildland firefighting, law enforcement, the military, construction, mining, logging, and more. A comprehensive evaluation may be needed to quantify the extent of physical stressors encountered in such occupations.

Analyzing the Equation

The lifting equation is used to calculate the recommended weight limit (RWL). The equation begins with a load constant (LC), 51 pounds (23 kg), that is decreased according to six task variables or multipliers (M): horizontal (HM), vertical (VM), distance (DM), asymmetrical (AM), frequency (FM), and coupling (CM). In this section we outline the steps in the procedure; for details of the process go to www.cdc.gov and look for the NIOSH lifting equation.

$$RWL = LC \times HM \times VM \times DM \times AM \times FM \times CM$$

HM: Horizontal distance is measured directly in front of the worker. If the distance is equal to or less than 10 inches (25.4 cm), the HM is 1.0. HM

Photo courtesy of U.S. Army

Based on its limitations, the NIOSH equation is not appropriate for most physically demanding occupations such as structural and wildland firefighting.

decreases as the horizontal distance increases. The value is 0.4 when H is 25 inches (63.5 cm) in front of the worker (see figure 15.1). For our example, we will place the load 18 inches (45.7 cm) in front of the worker, with a multiplier of 0.56; 51 pounds × 0.56 = 28.56 pounds.

VM: Vertical height is defined as the height of the hands above the floor. To determine VM, the deviation of V from 30 inches (76.2 cm) (knuckle height) is calculated. When V is 30 inches (76.2 cm), VM is 1.0. VM decreases with an increase or decrease in height from knuckle height. At floor level, VM is 0.78; 28.56 pounds × 0.78 = 22.77 pounds.

DM: The travel distance is defined as the vertical travel of the hands from the beginning to the end of the lift. For D equal to or less than 10 inches (25.4 cm), DM is 1.0. When D is 70 inches (177.8 cm), DM is 0.85. We'll use a distance of 35 inches (89 cm) with a multiplier of 0.87; 22.77 pounds × 0.87 = 19.81 pounds.

AM: The asymmetrical component refers to a lift that begins or ends outside the midsagittal plane, projecting directly in front of the body.

Asymmetrical lifts that occur on one side or the other should be avoided (see figure 15.2). The asymmetrical angle defines the degree to which the

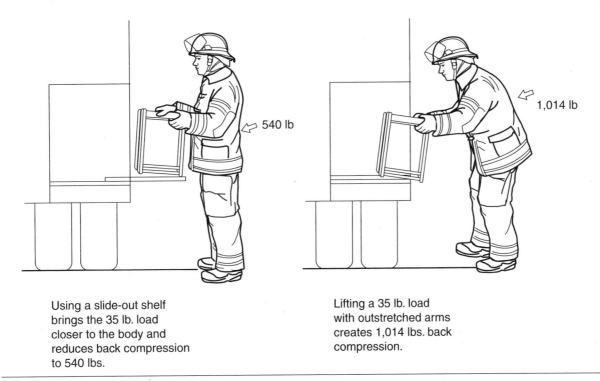

Using a slide-out shelf brings the 35 lb. load closer to the body and reduces back compression to 540 lbs.

Lifting a 35 lb. load with outstretched arms creates 1,014 lbs. back compression.

Figure 15.1   The effect of horizontal distance on back compression.

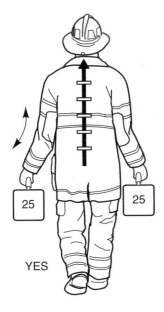
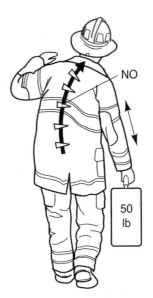

Spinal alignment maintained with balanced load. Arms in slight flexion to reduce stress on elbow.

Asymmetrical load forces a spinal curvature, pinching discs. Straight arm creates stress on elbow joint.

Figure 15.2   Asymmetrical load.

load lies away from the front of the body. When the load is directly in front of the body, the AM is 1.0. At 45 degrees, the AM is 0.86; 19.81 pounds × 0.86 = 17.04 pounds.

FM: The frequency multiplier is defined as the number of lifts per minute (frequency), the amount of time engaged in the lifting (duration), and the vertical height of the lift from the floor. For infrequent lifts, the FM is 1.0. Five lifts per minute for one to two hours yields an FM of 0.72; 17.04 pounds × 0.72 = 12.27 pounds.

CM: Coupling defines the hand-to-object grip. Good coupling reduces grasp forces and increases the acceptable weight for lifting. Good coupling rates a CM of 1.0, fair coupling rates a CM of 0.95, and poor coupling earns 0.9; 12.27 pounds × 0.95 = 11.65 pounds.

In this discussion the lifting equation reduced the load constant from 51 pounds (23.1 kg) to a recommended weight limit of 11.65 pounds (5.3 kg), a load that should be safe for most workers (95 percent). The equation identifies factors that increase the risk of lifting-related low back injuries. If the load to be lifted exceeds the RWL, and if the task cannot be redesigned, the relative stress of the job can be evaluated by calculating the lifting index (LI):

$$LI = Load\ weight\ /\ RWL$$

If the load weighs 25 pounds (11.3 kg) and the RWL is 11.65 pounds (5.3 kg), the LI exceeds 2. Lifting tasks with an LI greater than 1.0 pose an increased risk for lifting-related low back problems for a portion of the workforce. When the LI exceeds 3, nearly all workers will be at an increased risk of work-related injury. If the RWL is 30 pounds (13.6 kg), a load in excess of 90 pounds (40.8 kg) would pose that risk. Of course, there are other ways to reduce the risks of lifting and to improve the productivity of the workforce.

Selection and Training

The lifting equation is a compelling argument for establishing worker selection and maintenance standards. Many occupations require workers to lift and carry more than the load constant, 51 pounds (23.1 kg). How can they get the job done safely and efficiently? The answer is to use employee selection procedures and mandatory job-related fitness training. Employees who pass work capacity tests are able to perform potentially stressful tasks without significantly increasing their risk of work-related injury. Incumbent employees maintain muscular and aerobic capabilities by participating in work-related training. Ongoing muscular fitness training maintains muscle tone in abdominal and trunk muscles, thereby reducing the risk of low back injury. For example, chronic low back pain patients used lumbar extension exercises to improve muscle strength, endurance, vertebral bone mineral density, and joint mobility (Carpenter and Nelson 1999). Improvements occurred independent of diagnosis, were long-lasting, and appeared to reduce the use of the health care system. Low back strengthening shows promise for the reduction of industrial back injuries and their associated costs.

There is an association between poor fitness levels and increased injury rates in firefighters, military personnel, manufacturing workers, and manual material handlers. Resistance training improved material handling in the military (Williams, Rayson, and Jones 2002), for men and for women (Kramer et al. 2001). Combined muscular

---

CASE STUDY

Construction Loads

A field study investigated work-related causes of musculoskeletal disorders in the construction industry. Data were compiled on 340 workers in a variety of construction occupations. Scaffolders, bricklayers, and carpenters handled weights greater than 22 pounds (10 kg). With respect to lumbar disc (L5/S1), the scaffolders and bricklayers often exhibited excess pressure. Bricklaying required bent postures for 21 to 36 percent of daily work time. Painters, plumbers, and carpenters often worked in kneeling postures. The painters used overhead positions, whereas the bricklayers and scaffolders had high frequencies of material handling. The study showed that it is possible to rank different construction tasks with respect to load exposure, and that preventive measures are needed (Hartman and Fleischer 2005).

CASE STUDY
Training for Tree Planting

Tree planting has been shown to be a physically demanding occupation with an injury rate of approximately 12 percent, far in excess of the all-industry norm of 3.2 percent. A typical workday consists of planting upward of 300 trees per hour while covering 16 kilometers over difficult terrain. Seven to nine continuous hours are spent working with only a few minutes' break for food or rest, and workers display increased levels of stress hormones, significant body mass losses, and indicators of immune suppression. In view of the association between poor fitness levels and increased injury rates in firefighters, infantry soldiers, manufacturing facility employees, and manual material handlers, Roberts (2004) investigated the impact of a task-specific preseason fitness program on work productivity, injury rates, and changes in the level of selected biochemical markers of stress following work tasks.

Fourteen planters completed an eight-week preseason training program (TrG), with an additional 16 experience-matched planters acting as a control group (CG). There were no differences between the two groups in gender, age, height, weight, body composition, estimated $\dot{V}O_2$max, handgrip strength, or smoking behavior. The training program was an eight-week progressive design, with two high-intensity aerobic interval programs and one longer Fartlek program each week. The three resistance workouts were executed on alternate days to the aerobic sessions. Resistance training used elastic banding and 1 3/4-inch (4.4 cm) FlexBars to simulate planting movements with both concentrically and eccentrically loaded resistance. Planters were also provided with a small weighted sack (187 g for men and 165 g for women) for complex training of the wrist joints. Both groups were subsequently monitored during a monthlong planting contract.

Following training there was an increase in $\dot{V}O_2$max (from 47.2 to 55.7 ml · kg⁻¹ · min⁻¹) and handgrip strength (from 43.2 to 48.9 kg) in TrG planters. This group also demonstrated enhanced fatigue resistance by maintaining the same plant-

ing rate from morning to afternoon. In contrast, CG planters were significantly slower to plant 50 trees in the afternoon compared to the morning (see figure 15.3).

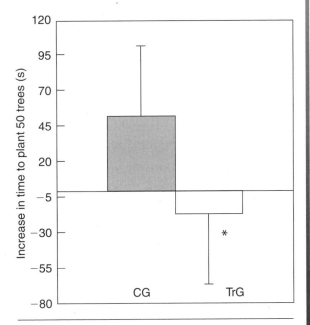

Figure 15.3 Tree planting. The difference in time to plant 50 trees is significantly longer in the afternoon compared to the morning.

* Significantly less change than CG planters.
Roberts 2006.

The TrG planters had higher overall productivity, planting more trees per day than the CG planters, which was weakly correlated to $\dot{V}O_2$max ($r = .34$). In addition, TrG planters performed significantly better in a 30-second neuromuscular coordination test. The most significant finding of the study was that the TrG planters suffered significantly fewer illness and injury events than the CG planters did, reporting only 1 event per planter compared to 2.4 events in the control group

Contributed by Delia Roberts, PhD, Selkirk College, British Columbia (2006).

and aerobic training improved material handling in women (Knapik 1997).

Before we conclude this chapter, let's review the things workers should do before lifting, during the

lift, and as they are lowering the load. The first thing to do, weeks in advance, is to develop job-related muscular and aerobic fitness. In addition to lifting muscles, workers should focus on trunk muscles and core training. The Canadian Centre for Occupational Health and Safety recommends the following (www.ccohs.ca):

Before lifting: Assess the weight of the load.

Can you lift it without risk?

Are mechanical aids available?

Get help for heavy loads.

Is the load free to move?

Is the path to destination free of obstacles?

Is the destination free of obstacles?

Are you warmed up and ready to lift?

Don't lift if you are not sure you can handle the load.

Lifting: Stand close to the load and face the direction in which you intend to move.

Use a wide stance to gain balance.

Be sure you have a good grip on the load.

Keep your arms straight.

Tighten your abdominal muscles.

Tuck your chin into your chest.

Initiate the lift with your body weight.

Use your legs instead of your back muscles.

Lift the load as close to your body as possible.

Lift smoothly without jerking.

Avoid twisting and side bending while lifting.

Lowering the load: Take a wide stance with one foot in front of the other.

Keep the load close to your body.

Keep your back straight.

Bend your hips and knees.

Set the load down onto the ground.

Stand up smoothly, easing your muscles.

Avoid a jerky release.

Now it's time to take a break from this heavy lifting and read the final chapter on legal issues.

Summary

Lifting limitations have been studied as a way to reduce workplace injuries. Lifting guidelines have been established using psychophysical, biomechanical, and physiological standards. The revised NIOSH lifting equation has used these techniques to determine the recommended weight limit for two-handed lifts. Unfortunately, the limitations of the equation reduce its usefulness for physically demanding occupations. The equation attempts to determine lifting limits for the majority of workers, making it even less useful for hard work. Firefighters couldn't wear turnout gear and SCBA and then carry heavy loads up several flights of steps. Wildland firefighters couldn't carry pumps and hose rolls and climb hills while working on uneven ground in high ambient temperatures. And military personnel couldn't carry heavy packs and weapons, or load and fire heavy shells. Most of those loads exceed 51 pounds (23.1 kg), the NIOSH load constant.

How can lifting injuries be avoided in demanding occupations? Alternatives to the lifting guidelines include job-related work selection tests and career-long work-related fitness programs. Employers in demanding occupations should carefully select and train employees to avoid low back and other injuries. Carefully selected workers can perform potentially stressful tasks that would be hazardous to the workforce at large. But that protection will diminish without participation in a training program. Poor fitness levels are associated with increased injury rates in physically demanding occupations. There is a saying: Don't play sports to get in shape; get in shape to play sports. The same could be said for physically demanding occupations.

Chapter 16
Legal Issues

American Color photography

Since 1964 the passage of landmark legislation prohibiting discrimination in the workplace and the development of the Uniform Guidelines on Employee Selection Procedures in the United States have transformed the strategies employed for personnel selection and ongoing employment. The guidelines define adverse impact as a pass rate below 80 percent (four-fifths rule) of the pass rate of the comparison group. Adverse impact applies when there is a substantially different rate of selection in hiring, promotion, or other employment decisions that works to the disadvantage of members of a given gender or a racial or ethnic group. If adverse impact exists, the employer must provide evidence that the test or requirement is valid, that it is related to the job in question, that its use is essential to the safe and efficient operation of the business, and that there are no alternative procedures available that are equally valid to achieve the business objectives with less adverse impact (for more information, see www.uniformguidelines.com).

With over 70 years of experience in work physiology, we have observed the passage of laws, growing numbers of women in the workforce, and increasing legal challenges and court cases on preemployment testing and job performance standards. We have created job-related selection tests and physical maintenance programs. And we have been called on to serve as consultants and expert witnesses in a wide range of arbitrations and legal challenges. We are not lawyers, but we know a lot about work physiology, health, safety, and performance. This chapter shows how some legal decisions have, in our opinion, misinterpreted or misrepresented established facts in work physiology, thereby diminishing the health, safety, and performance of those working in physically demanding occupations.

This chapter will help you do the following:

- Understand the reasons for legal challenges.
- Acknowledge the unintended consequences of legal challenges on workers, the workforce, productivity, and health.
- Compare favorable and unfavorable court decisions.
- Consider an alternative approach to litigation.

Legal Challenges

Legal disputes based on preemployment testing or the maintenance of work capacity date back to the Civil Rights Act of 1964 and subsequent U.S. employment laws, including the Age Discrimination in Employment Act of 1967 (ADEA) and the Americans with Disabilities Act of 1990 (ADA). A review of 44 court cases dealing with physical standards and preemployment testing found that 34 involved height and weight standards and 10 dealt with work capacity testing (Hogan and Quigley 1986). In the cases involving recruit testing for fire and police positions, 8 of 10 cases were decided in favor of the job applicants. Who are these plaintiffs, and what are the consequences of the cases?

Plaintiffs

The plaintiff is the person or group that brings a suit in court. Preemployment discrimination cases are often brought by women when tests are shown to have adverse impact. That was true in the *Lanning v. SEPTA* case (2002), *United States v. City of Erie* (2005), and the Canadian Supreme Court case (*British Columbia [Public Service Employee Relations Commission] v. British Columbia Government Service Employees' Union* 1999). Failure of one or more women on the test leads to hearings or arbitration, and possibly to court.

In the *SEPTA* and *Erie* cases, the U.S. Department of Justice (DOJ) was involved in the suit. In the *Erie* case, the DOJ sued on behalf of more than 100 women who had failed the often-changed physical test between 1996 and 2002. The challenged test was dropped after 2002 and replaced with one scaled for age and gender. The Department of Justice has responsibility for public employ-

ers, whereas the Equal Employment Opportunity Commission (EEOC) is responsible for private sector employers. In the Canadian case, the suit was joined by the B.C. Human Rights Commission, the Women's Legal Education and Action Fund, the Disabled Women's Network of Canada, and the Canadian Labor Congress (see box).

‖‖‖‖‖ SPECIAL INTEREST ‖‖‖‖‖ GROUPS

The Canadian Supreme Court case illustrates the fact that legal challenges are frequently joined by groups with an interest in the outcome of the proceedings. The groups are often advocates for women's rights, for older workers, or for the disabled. Unfortunately, few individuals or groups intervene on behalf of health, safety, and performance. Perhaps the only advocates for that agenda are the defendant's attorney and expert witnesses.

Expert Witnesses

If scientific, technical, or other specialized knowledge will assist the trier of fact to understand the evidence or to determine a fact in issue, a witness qualified as an expert by knowledge, skill, experience, training, or education, may testify in the form of an opinion (Fundamental Rule of Evidence Rule 702). The court may appoint any expert witnesses agreed on by the parties and may appoint expert witnesses of its own selection (FRE Rule 706). A typical employment discrimination case involves one or more expert witnesses on each side (Bronstein 1993). The witnesses are chosen based on their credentials and experience and on their position relative to the case. Some "experts" have never developed or validated a test or designed a job-related fitness program. They may be experts in research design, statistics, or even exercise physiology. Some expert witnesses are always arguing for easier tests and lower standards of performance. Others fight to maintain valid job-related standards.

The *Lanning v. SEPTA* case (2000) involved several expert witnesses on each side. The court affirmed the need for an aerobic capacity of 42.5

Expert witnesses are chosen based on their credentials and experience and on their position relative to the case. Some expert witnesses argue for easier tests and lower performance standards, whereas others fight to maintain valid job-related standards.

(ml · kg⁻¹ · min⁻¹) as the minimum level required to perform essential transit officer tasks, and that SEPTA had met its burden of establishing the business necessity of its aerobic capacity standard. SEPTA offered Dr. Paul Davis' calculation of the aerobic capacity required to perform essential tasks; Dr. Siskin's arrest rate studies, including analysis of performance differences between those officers always at 42.5 versus those never at 42.5; Dr. Moffatt's study on work output decrements associated with aerobic capacities below SEPTA's cutoff point; and the most recent study that demonstrated that the cutoff point was already at the minimum. The court decided that a standard below 42.5 would result in officers being unable to successfully perform the job of transit police officer.

80 Percent Rule

The arbitrary designation of 80 percent as a test of adverse impact opens the door for legal action. Where did the 80 percent, or four-fifths, rule come from? Untrained

women have been shown to have approximately half of the upper-body and two-thirds of the lower-body strength of men (Miller et al. 1993). Data on wildland firefighters confirm those percentages, with women exhibiting 47 percent of the upper-body and 62 percent of the lower-body strength of men (Sharkey 1981). Data for women in the U.S. military indicate that women have 60 percent of the upper-body strength and 67 percent of the lower-body strength of men (Sharp 1994). These percentages almost guarantee adverse impact in a test that demands upper-body strength.

Muscular endurance and aerobic fitness are less likely to trigger adverse impact. Muscular endurance depends on the test and the opportunity to train. For aerobic fitness, women average about 85 percent of men. In contrast to evidence accepted in the Canadian Supreme Court decision (British Columbia 1999), women are just as able as men to improve aerobic fitness with training.

Lance Cpl. Christiana Fowlkes, a database clerk with Headquarters and Service Company, Headquarters Battalion, 2nd Marine Division, does the flex arm hang, one of the required parts of her physical fitness test. Untrained women have been shown to have approximately half of the upper-body strength of men, which almost guarantees adverse impact in a test that demands upper-body strength.

IIIII **GENDER AND FITNESS** IIIII

Before puberty, boys and girls differ little in aerobic fitness, but from then on girls fall behind. Young women average 85 percent of the aerobic fitness of young men, depending on their level of activity. But highly trained young female endurance athletes are only 10 percent below elite males in $\dot{V}O_2$max and performance times. Until the 1970s, women were not allowed to compete in races longer than one-half mile (0.8 km). Overprotective or prejudiced officials worried that frail females couldn't stand the strain. Today women run marathons and 100-mile (160 km) races; compete in the Ironman triathlon; and swim, ski, and cycle prodigious distances. A woman led much of the 1994 high-altitude Leadville 100 Mile Trail Run until she was passed by a male Tarahumara Indian from Mexico. We've learned that women are well suited for fat-burning endurance events and that some tolerate heat, cold, and other indignities as well or better than men. When properly trained, women's endurance performances are 10 percent slower than men's, which is not enough to trigger adverse impact or invalidate a job-related aerobic test.

The maintenance of fitness and work capacity has been challenged by older employees and those unwilling or unable to meet the standards because of lack of training or excess body weight. For example, an arbitrator in Oregon found that a fire department violated the employment agreement by ordering firefighters to perform an endurance test (*Klamath Falls Fire District v. KF Firefighters*, 1996). A U.S. federal court rejected a physical agility test for an Illinois city on the basis of insufficient test validity (*Thomas v. City of Evanston*, 1983). On the other hand, a federal court upheld physical and training standards retroactively imposed on experienced volunteer firefighters (*Alexandria Volunteer Fire Department v. Rule*, 1984). A court ruled that an Iowa city could adopt a physical fitness test for the retention of firefighters, but that those who fail must be allowed to challenge the validity of the test (*Des Moines v. Civil Service Commission*, 1995). The laws provide the opportunity to challenge a test, but what are the consequences?

Unintended Consequences

When a court rejects a test or arbitrarily lowers test or job maintenance standards, what are the consequences? When tests are rejected, employers are forced to accept less qualified applicants. The same is true when an employer is denied the opportunity to enforce ongoing fitness standards. The employer or taxpayers are forced to settle for minimum performance and an increased risk of injury and chronic illness. Fit police officers have a superior arrest rate. Fit firefighters accomplish more in less time. Fit workers are less prone to overexertion injuries, and they are much less likely to suffer from overweight, diabetes, heart disease, and some cancers. Physically demanding law enforcement and firefighting positions were created to protect and defend public health and

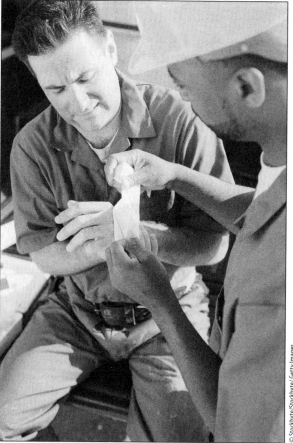

When an employer lowers test standards or is denied the opportunity to enforce ongoing fitness standards, it is forced to settle for minimum performance and an increased risk of injury among workers.

© Stockbyte/Stockbyte/Getty Images

safety. Yet we seem willing to ignore performance and physiological facts to accomplish what some have called social engineering.

What are the consequences for those who secure a position via legal challenge? Are lower standards beneficial to women, older workers, or overweight employees? The short-term answer is probably yes, because the employee has a job he or she couldn't get with truly job-related standards. The long-term answer could be no if the employee isn't promoted, gets injured, or develops a chronic illness. An unintended consequence of low hiring and job maintenance standards is an adverse effect on all employees. Fit workers regress to a lower level, and less-than-fit workers never achieve their potential. For example, the opening of distance races to women allowed their endurance to emerge. Maintenance of high-performance job standards does the same. Moreover, high standards attract higher-quality applicants, regardless of age or gender.

A job worth having is one worth training for. Are women aided by lower standards? No, and neither are the men they work with. Thousands of people have the capacity to succeed in physically demanding occupations. The Uniform Guidelines on Employee Selection Procedures do not reject tests with adverse impact. Instead, they call for a validation that proves the relationship of the test to the job. The courts seldom require evidence that candidates have trained for a test or have trained to maintain job standards. In a society plagued with an epidemic of overweight and obesity and diminishing physical activity and fitness, the courts seem willing to accept that situation as the norm for the population. That is a consequence society should not accept and cannot afford.

Court Decisions

Court decisions have been both favorable and unfavorable for the establishment of job-related entry standards and ongoing fitness requirements. Some lawyers suggest that federal judges with lifetime appointments are more likely to give standards a chance, and that elected state judges are more susceptible to pressure from intervening groups. What you will see is that decisions fall on both sides of the issue. In this section we present cases on police and fire personnel, occupations in which the majority of physical fitness requirements and maintenance standards occur.

Unfavorable Decisions

An unfavorable decision, in our view, is one that rejects a test, lowers standards, or eliminates ongoing fitness requirements. Note that some cases refer to an agility test, which is an outdated term for a job-related or fitness test.

United States v. City of Erie, 2005: The U.S. Department of Justice sues Erie, Pennsylvania, over an alleged gender-biased entry test. According to the court, the city failed to meet its burden of proof with respect to the test (see case study).

CASE STUDY
The *Erie* Case

The Erie test required applicants to run a 220-yard (201 m) obstacle course in 60 seconds and then perform 17 push-ups and 13 sit-ups, each in 15-second sessions, for a total of 90 seconds. The time and number of push-ups and sit-ups were based on the average performance of a group of incumbents. The aggregate pass rate for men was 71 percent, and the rate for women was 13 percent. Although the city's expert witness concluded that the test was too easy for the demands of the job, a view shared by incumbent officers, the court ruled against the city.

This was the first application of the *Lanning (SEPTA)* case in the Third Circuit. The centerpiece of the judgment was the court's stringent interpretation of the *Lanning* requirement that a disparate test may only be validated if its passing standard represents "the minimum qualifications for successful performance of the job." The judge seemed to be looking for an inarguable statistical association of the test's passing score with success as an Erie police officer. Rather than proceeding to a damage phase, the Justice Department and Erie entered into a settlement that provided for the possibility of hiring up to five women who had previously failed the test, so long as they could pass the updated (gender-adjusted) hiring standards. The settlement also included some payment to women determined eligible for "back pay" (Personal communication from G. Villella, deputy solicitor, City of Erie, 2006).

Klamath Falls Fire District v. KF Firefighters, 1996: An Oregon arbitrator finds that a fire district violated the employment agreement by ordering firefighters to perform an endurance test.

White v. Village of Homewood, 1993: A firefighter applicant who was injured while taking a preemployment agility test could recover against the municipality for any negligence.

Legault v. Russo, 1994: A U.S. federal court enjoins the use of a biased firefighter agility test and orders employment of a rejected woman applicant.

United States v. New Castle, Delaware, 1984: The U.S. Justice Department obtains a consent decree against Delaware City for the rejection of female applicants who failed physical tests.

Thomas v. City of Evanston, 1983: A U.S. federal court rejects a physical agility test used by an Illinois police department for insufficient validity.

Patrolmen's Benevolent Association v. Township of East Brunswick, N.J., 1984: The court ruled that lowering physical agility test scores for all does not discriminate against male applicants.

Meridian v. Firefighters Association of Michigan, 1996: Management had a duty to bargain with a firefighters' union before implementing a mandatory agility test and allowing discipline to be imposed on those who declined to participate.

Utica Professional Firefighters v. City of Utica, N.Y., 1999: Unilaterally implementing a requirement that firefighters undergo and pass respiratory fitness examinations was mandatorily negotiable.

Favorable Decisions

A favorable decision is one that retains a test, maintains standards, or allows ongoing fitness requirements.

Township of Bridgewater, N.J. v. P.B.A. Local 174, 1984: A New Jersey appellate court upholds mandatory fitness tests and decrees that they are not a proper subject for bargaining or arbitration.

Pentagon Force Protection Agency v. Fraternal Order of Police DPS Labor Committee, 2004: A U.S. Federal Labor Relations regional office concludes that management, in creating a physical fitness program for Pentagon police officers, was not required to bargain over a grandfather clause or the creation of a Medical Review and Physical Fitness Board.

Smith v. Des Moines, 1996: A U.S. federal appeals court affirms the termination of a fire captain who failed a spirometer test with a $\dot{V}O_2$max of 28.9. (A

$\dot{V}O_2$max requirement of 33.5 was upheld in an ADA and ADEA challenge.)

Des Moines v. Civil Service Commission, 1995: An Iowa city could adopt an annual physical fitness test for the retention of firefighters, but those who fail it must be allowed to challenge the validity of the test.

EEOC Enforcement Guidance on Preemployment Inquiries under the ADA, 1995: The EEOC's ADA guidelines allow preoffer agility and physical fitness tests provided they are job related and necessary (www.eeoc.gov/policy/docs/preemp.html).

Zamlen v. City of Cleveland, 1988: A U.S. federal court upholds physical agility tests for firefighter applicants; the court concludes that the anaerobic component is more important than aerobic component.

Ferrante v. Niagara County, 1990: A U.S. appellate court upholds the rejection of a police applicant who failed the running test. He was not entitled to substitute a different test even though another candidate was allowed to do so.

Eberle v. State of Missouri Department of Corrections, 1989: A U.S. appellate court says an injunction is not a proper way to contest the validity of new physical endurance standards for corrections officers who are selected for or retained in an emergency response unit.

United States v. City of Wichita Falls, 1988: A federal court upholds the use of a timed wall climb, dummy drag, and agility movements in a test for police applicants in Texas. The test was ruled as job related even if it had a disparate impact on female applicants.

United Paramedics of LA v. City of Los Angeles, 1991: A federal court upholds the Los Angeles Fire Department weight standards challenged by paramedics.

Pierce v. Franklin Electric Co., 1987: An employee may be discharged for physical inability to do work.

Alexandria Volunteer Fire Department v. Rule, 1984: A U.S. federal court upholds physical and training standards retroactively imposed on experienced volunteer firefighters.

Costs of Litigation

The *SEPTA* and *St. Paul* cases took several years; a number of lawyers; expert witnesses; and the time and effort of federal, state, and city employees to be resolved. In both cases the considerable financial cost was borne by federal, state, and city taxpayers. If you are thoroughly confused by the inconsistent application of the law, you are not alone. Law review articles have explored aspects of employment testing, including the provision that tests must be job related and consistent with business necessity.

Lanning v. SEPTA, 1998: The Third Circuit overturns a 1.5-mile (2.4 km), 12-minute run requirement that disqualified most women police officer candidates. The court ruled that the test was unrelated to overall job performance. The U.S. Supreme Court denies a review.

Lanning v. SEPTA, 2000: The U.S. Justice Department drops its opposition to a timed run used to screen transit police officers in Philadelphia. The ruling upheld a transit police force requirement that all applicants run 1.5 miles (2.4 km) in 12 minutes or less. The trial court ruled, for the second time, that the test did not unlawfully discriminate against women.

Along the way the case was remanded to hearing, the argument being that there had to be empirical evidence to support a minimal level of fitness and that that level could not be based solely on expert opinion. The data collection cost hundreds of thousands of dollars, to validate the obvious.

Lanning v. SEPTA, 2002: A divided federal appeals court, in its second opinion, upholds a 1.5-mile (2.4 km) run requirement that disqualifies many women candidates from consideration for the position of transit police officer.

RULE OF LAW

The United States is a nation founded on the rule of law. We have ordinances, codes, regulations, and public law. The U.S. Congress has the authority to establish the law of the land. These public laws are called statutory law. Lesser bodies such as states and municipalities may pass codes, ordinances, or regulations. Likewise, Congress may pass general regulatory laws empowering various agencies of the federal government to enforce laws with their own regulations (e.g., the Occupational Safety and Health Administration [OSHA] with workplace violations and the EEOC with enforcement of the ADA). the term *codified law* is used to organize statutes under similar subject matter headings.

Common law, inherited from Great Britain, may have relevance in cases in which a statute has not been written to cover a certain set of circumstances. Common law does have precedence and can be cited with authority. Civil law distinguishes itself from criminal law. In both cases there is a defendant. However, in the criminal matter there is the prosecution, whereas in civil matters plaintiffs are the parties seeking relief, usually money, as a form of making "whole" their damages.

Case law may have precedent when there have been a number of lawsuits over the interpretation of a statutory law. Ultimately, the highest court in the land has the power to decide whether a law is consistent with the Constitution of the United States. Attorneys will cite case law as a method of showing that similar cases with similar facts have been decided in their favor.

In the United States there are 12 judicial circuits with a total of 94 district courts. A decision made at the local district court may or may not be binding on another district. Similarly, a decision by one of the 13 appellate courts may be narrowly applied to a similar set of circumstances within that appellate court's geographic area.

The Uniform Guidelines on Employee Selection Procedures is a consensus document promulgated by four federal agencies (the Equal Employment Opportunity Commission, the Department of Labor, the Department of Justice, and the Office of Personnel Management). It closely follows the tenets for the validation of written tests. First published in 1978, this consensus document is a blueprint for defining the pathway to validity for tests used in employment settings. Heavily influenced by the industrial and organizational psychology profession, it was never envisioned to be used for the design of physical ability tests. This oversight has resulted in attempts to engineer tests around the capabilities of women rather than the requirements of the job. Although the Uniform Guidelines are not public law or even a regulation, great deference has been given to them.

Title VII of the Civil Rights Act of 1991 was passed by Congress to address lapses in Supreme Court decisions. Despite these efforts to make clear the intentions of Congress in creating a level playing field for employment opportunity, interpretation of Title VII is still a hodgepodge with decisions scattered about the political landscape.

(Contributed by Saul Krenzel, LLD, Defense Council, *Lanning v. SEPTA*)

CASE STUDY
Civil Rights Act of 1991

Title VII of this act provides for job-relatedness and business necessity defenses to adverse impact claims. According to Hollar (2000), over the last 20 years at least 12 published cases evaluated claims asserted by females seeking to invalidate physical ability tests. The decisions diverged widely, splitting almost in half on findings of liability, using numerous standards for applying the business necessity defense. Hollar stated that the lack of clarity in the area makes it hard both for employers to design valid tests and for applicants to successfully press claims of genuine gender discrimination. He argued that the courts and Congress have struggled with Supreme Court precedent formulating the business necessity defense because of the failure to differentiate between the strands of analysis that have gone into that defense. He proposed that job-relatedness and business necessity should be treated as separate requirements, both of which must be met for an employer to successfully defend the validity of its test. (See also Sarno 2003.)

Human Rights Commission, State of Minnesota v. St. Paul Fire Department, 1989. O.A.H. Docket No. 8-1700-3224-2: A preservice agility test for the St. Paul Fire Department was ruled discriminatory and unsupported by realistic job requirements. In this case, the expert witness for the plaintiff had no experience in the design of physical ability tests, yet he opined that firefighters self-pace at the scene of a fire, thereby negating the requirement for a high level of fitness.

Human Rights Commission, State of Minnesota v. St. Paul Fire Department, 1989. O.A.H. Docket No. 8-1700-3224-2: The St. Paul Fire Department settles a sex discrimination suit and formally adopts the Davis Criterion Task Test.

The final section of this chapter proposes an alternative approach to this confusing and costly process.

Alternative to Litigation

Over the past 30 years, hundreds of legal challenges to job-related fitness standards have consumed millions of tax dollars. Litigation or fear of suits over age, gender, or disability has paralyzed some employers to the point where the path of least resistance results in no standards at all. We have presented cases that affirmed and rejected job-related employment tests and ongoing physical standards. There must be a better way.

There are many hundreds of city, county, and rural fire departments and almost as many city and county police departments in the United States. The fire organizations follow national guidelines for training and operations. Law enforcement groups use their own national guidelines. Why then do we need separate tests and standards for each city or county? We recommend developing an alternative approach to solve hiring and job maintenance issues. It begins with a collective approach to determine the essential functions of the job, uses independent expert witnesses, verifies assertions with peer review, and concludes with a consensus decision.

Collective Approach

Should every city and county fire and police department have to develop and validate its own test? Why don't national organizations validate standards for firefighters and for law enforcement personnel? The tests could have a basic core of essential functions with additional elements for use in unusual environments (e.g., a tall building element for city firefighters, scaling a 6-foot [182 cm] wall for southwestern law enforcement). The test could be validated on a national sample and approved for use on all recruits.

For example, the National Fire Protection Association (NFPA) could use its standards process to develop and validate the test. Typically the process involves firefighters, fire managers, subject matter specialists, technical help, and, when appropriate, representatives from industry (e.g., clothing or equipment manufacturers). An independent organization such as the American College of Sports Medicine (ACSM) could be contracted to review the test development process and the sampling design and ensure the accuracy of factual assertions (e.g., women's' ability to improve with training). Once the test has been developed, it could be pilot tested on recruits in select locations. If adverse impact is identified, the test could be reviewed to see whether adverse impact could be reduced while maintaining the validity of the test. A similar approach could be used in law enforcement.

Firefighter Standards

Several years ago the NFPA convened a group to develop fitness standards for incumbent firefighters. The working group met regularly to develop job-related fitness standards. In the program overview it stated:

"Fire fighting is one of the most physically demanding and dangerous occupations. It is the responsibility of the fire department to maintain a work force that is prepared to handle the physical rigors of the job safely and proficiently."

"All current firefighters should cooperate, participate, and comply with the provisions of the fire department's occupational safety and health program."

"The fire department should establish a physical fitness and health enhancement program that promotes optimal physical preparation for the job of fire fighting. The program should include physical fitness and performance requirements that prepare and maintain fire fighters in the best physical condition they can achieve throughout the length of their service as fire fighters. At least as much emphasis should be placed on the maintenance of physical fitness of the individual fire fighter as on the preventive maintenance of facilities, vehicles, and equipment."

In addition to physical conditioning and physical fitness assessment, the proposed program included a physical performance assessment for the initial selection of recruits and the annual evaluation of incumbents. The assessment included a stair climb with load, hoisting, forcible entry and ventilation, hose advance, a carry evolution, and a victim rescue. Although progress toward meaningful standards seemed possible, it was eventually defeated by special interest groups, including the firefighters' union and a women's advocacy organization.

After several years, NFPA failed to achieve consensus on mandatory fitness standards and discontinued its efforts. The organization then developed NFPA 1583, Standard on Health-Related Fitness Programs for Fire Fighters (2000), to enable members to develop and maintain an appropriate level of fitness to safely perform their assigned functions. The fitness levels in the program are to be based on fitness standards determined by the fire department physician that reflect the individual's assigned functions and activities and that are intended to reduce the probability and severity of occupational injuries and illnesses. Unfortunately, there are no mandatory nationwide standards and few mandatory fitness or performance requirements.

The NFPA committee met several times a year during the three-year effort to establish meaningful standards. Most committee members traveled at department or organization expense and were paid during their participation. A nationally recognized subject matter expert attended at his own expense because he recognized the importance of fitness and performance standards to the health and safety of firefighters. He continues these efforts as the developer and chief executive officer of the Firefighter Combat Challenge.

Law Enforcement Standards

In an attempt to provide guidance to the law enforcement community, a working group consisting of members from the Major Cities Chiefs, National Executive Institute Associates, and the FBI was assembled in Quantico, Virginia, in 1993. Its mission was to lay out guidelines and recommendations for establishing standards for physical performance in light of case law and the anticipated impact of the ADA and the Civil Rights Act of 1991. The work product titled "Physical Fitness Testing in Law Enforcement" sets forth the notion that performance-based testing represents the most defensible approach to identifying individuals for positions within the profession, and that the identification of cutoff points and retention standards should be carefully considered and thoughtfully designed. The preponderance of opinion of the subject matter experts on the Major Cities Chiefs' panel (22 members including police attorneys, personnel specialists, test construction experts, and law enforcement officers) supported the use of gender- and age-neutral testing that included samples of work performance and a stamina component to test for cardiorespiratory fitness. This document also included an overview of case law with special emphasis on public safety implications.

IIIIIIIIIIIIIIII ESSENTIAL FUNCTIONS: LAW ENFORCEMENT IIIIIIIIIIIIIIIIII

According to the U.S. Bureau of Justice Statistics, there are over 800,000 law enforcement officers in the United States, representing 17,874 state and local agencies and nearly 100 federal agencies. With this diversity, is there a commonality that defines the essential functions of the job? Does every department have to research, design, and implement physical performance standards that are unique to its specific mission? In reality, law enforcement agencies are distinguished by the requirement to "serve and protect"; specifically, to invoke the power of arrest. Noteworthy is the common requirement across the entire spectrum of law enforcement to carry a weapon and apply lethal force.

A cursory inspection of the curriculum for certification as a police officer in any setting—from the FBI Academy or FLETC (Federal Law Enforcement Training Center) to the smallest regional criminal justice academy—contains a program of instruction (POI) covering the use of firearms, less-than-lethal force, defensive and offensive tactics, and physical training. This commonality of critical learning objectives illustrates the point that core competencies already exist.

Central to the body of knowledge that has developed since the 1960s is the force continuum, the escalation of force required in securing the compliance of a suspect, violator, or perpetrator. When a suspect decides to flee the scene or engage an officer in a physical struggle, it is an issue of officer versus the subject.

Tests of muscular fitness and power are associated with success in law enforcement tasks. In case there is any remaining doubt about the occurrence of such tasks as scaling walls, entering windows, dragging or lifting bodies, crawling under or over an assortment of obstacles, and restraining and obtaining compliance in resisting perpetrators, data collected by Paul Davis, Cunningham (2006), and many others consistently reveal a high level of agreement on the performance of these tasks.

After hundreds of job task analyses in a variety of organizations, there is a clear trend toward the identification of a core set of essential functions that are common to law enforcement officers throughout the United States. A similar set of essential functions exists for structural firefighting. For a discussion of essential functions of law enforcement, see appendix H.

IIIIIIIIIII WILDLAND FIRE IIIIIIIIIIII

The collective approach has been employed successfully by the five U.S. federal agencies involved in wildland firefighting. After extensive laboratory and field testing, a test was administered to a sample of 320 and then to 5,000 firefighters. The work capacity tests (pack test, field test, and walk test) were approved by fire leadership in each of the agencies. In addition to the five federal agencies, the tests are used by many state fire organizations, county and rural departments, Province of British Columbia firefighters, and several states in Australia.

Independent Experts

Independent experts, those with considerable experience in the field, could be asked to contribute to the test development process. This could minimize the "battle of the experts," in which some earn as much as $500 an hour of taxpayer dollars. The experts would not work for one side or the other; they would endeavor to guide the process. Experts in work physiology would ensure that the job standards are safe and within the reach of people who are willing to train. Experts in occupational medicine would determine related medical requirements. If adverse impact emerges, the experts could evaluate the effects of alternative approaches. Once a test has been developed, validated, cross-validated, and tested on sample recruits, it is ready to be considered for adoption.

Consensus

When the process (a test or ongoing fitness standards) has been tested and approved in field studies, it can be presented to the organization's decision-making bodies. In the case of the NFPA, that could be the appropriate technical committee and the Standards Council. Once approved, it becomes the standard throughout the country, saving individual departments the costs of test development and the risks of legal challenge. Unions interested in the health and safety of their members should favor this process. Failure to have mandatory standards contributes to a substantial risk of injury and illness and an unacceptable level of cardiovascular disease.

Summary

This chapter began with a discussion of some reasons for legal challenges of preemployment tests and incumbent fitness standards. Adverse impact on women is the most likely reason for challenging tests. Incumbents challenge fitness standards when they can't meet or refuse to meet the designated level of fitness or job performance. The challenges often lead to unintended consequences, including lower standards or the elimination of meaningful standards. Other unintended consequences are adverse effects on health, safety, job performance, and public safety.

An examination of favorable and unfavorable court decisions reveals a disturbing lack of continuity. Sometimes one court affirms a test and the appeals court rejects it, or vice versa. Court challenges are time-consuming and expensive. We suggest an alternative to litigation that includes a collective approach and independent expert witnesses and concludes with a consensus decision. Although this approach is not always successful, it has the potential to save many thousands of dollars while satisfying the interests of the parties involved.

This book has defined the dimensions of hard work and described the characteristics of the workforce. When men and women go to work today, few are required to engage in arduous muscular effort. And yet, physically demanding jobs still exist in firefighting, law enforcement, the military, construction, mining, forestry, and agriculture. It is clear that many recruits have failed to develop the capacity to perform physically demanding occupations. Some applicants lack endurance, some lack strength, and some lack the stamina and strength required to perform prolonged hard work safely and effectively. Very few applicants are fit enough to perform throughout a career without a commitment to a job-related fitness program.

Appendix A

Physical Ability Testing:
Conflicts, Conundrums, and Consequences

Paul O. Davis, PhD • First Responder Institute

Abstract

The physical ability testing (PAT) landscape for the hiring and retention of police officers is pitted with land mines of conflicting objectives. And, although there is virtual unanimity of the need for fit officers, police departments range in their practices from no standards at all to some semblance of physical fitness standards, typically based on normative data derived from clients of the Cooper Institute. Routinely absent is the evidence of comprehensive PAT entrance and retention standards that are tied directly to the specific skills or abilities demanded by the job. To be sure, formal job task analyses (JTAs) (McCormick 1979) support the requirement for officers who can successfully perform a range of activities that frequently require the pursuit and apprehension of resisting and fleeing felons. However, fear of litigation, grievances from employee associations, lack of funding, pressure from special interest groups, intrusions by elected officials, and lack of leadership characterize the underpinnings for failure to confront the implementation of hiring or retention standards. The courts have been of little assistance in ferreting out a clear pathway for departments that are interested in improving and maintaining the quality of their workforce. Presented here is an examination of Paul Davis' 10 Employment Axioms with support from published research and recommendations for a fresh start to improving the quality of the law enforcement workforce.

Background

For the past 30 years I have been involved in conducting physiological research on arduous, critical, and physically demanding occupations. In a first-of-a-kind research study, my colleagues and I have explored relationships among the varying levels of fitness that were demonstrated to have a hierarchical or ascendancy effect on the performance of arduous occupational tasks (Davis, Dotson, and Santa Maria 1982). Concurrently, I have tendered over 60 consultations or appearances for expert testimony in employment opportunity lawsuits. In an attempt to codify the relevant scientific findings and practical experience, this paper presents the basis and arguments in support of the 10 Employment Axioms. Reinforcing these axioms, evidence and testimony will be proffered from two recent employment opportunity lawsuits (*Lanning v. SEPTA* [1998, 2002]; *United States v. City of Erie* [2005]), both heard in Pennsylvania federal courts with similar facts, evidence, and expert witnesses. The outcomes are diversely different and a direct reflection of the district court judges' understanding of the importance of physically capable officers. This treatise is intended to provide a road map to a better understanding of the complexities of employee selection and how to balance the law enforcement mission and safety with society's interests and employment diversity.

Davis' 10 Employment Axioms Within Law Enforcement:

1. Employment in law enforcement is not a right; it is a privilege.

2. Although largely sedentary, the job does have, at times, profoundly demanding physical requirements.

3. The physical capabilities of the criminal element define the activities and the response of the law enforcement officer.

4. The physically fit officer has a greater probability of success than his or her less fit counterpart (i.e., more is better).

5. Physical fitness is an attribute that, if not maintained, degrades over time.

6. There are no prohibitions against applicants or incumbents improving their physical fitness commensurate with the requirements of the job.

7. No physical ability test is ever perfect.

8. There is no such thing as a physical ability test that does not have adverse impact.

9. The fact that a test has an adverse impact does not make it illegal.

10. There can be no physical fitness without a physical training program.

1. Employment in law enforcement is not a right; it is a privilege.

In our pursuit of diversity or other socially driven recruiting objectives, we sometimes forget that not everyone is suited to this occupation (as addressed by the Americans with Disabilities Act of 1990). Although we all believe in the U.S. Constitution's guarantee of equal opportunity, it does not guarantee equal results (*Personnel Administrator of Massachusetts v. Feeney,* 1979). One remarkable ideal and aspiration of this country is to encourage all people to go as far as their natural capabilities will take them. In a letter to William Herdon in 1848, Abraham Lincoln remarked, "The way for a young man to rise is to improve himself every way he can, never suspecting that anybody wishes to hinder him."

By contrast, adjusting the playing field to accommodate or give special treatment based on age, race, sex, or country of origin is specifically proscribed in the Civil Rights Act of 1991. As a consequence, you cannot have separate standards based on "identity codes" for the same job. The job requirements are independent of the person who is performing the job. Although this concept should seem straightforward, social engineering sometimes attempts to "adjust" standards to accommodate differences, thereby making the playing field "fair." This belies the fact that fairness is not a scientific or statistical concept and that, like beauty, fairness exists in the eyes of the beholder.

2. Although largely sedentary, the job does have, at times, profoundly demanding physical requirements.

Often heard is the comment, "Really, how often does a police officer have to chase someone?" In answer to this question, consider the parallel to the position of a lifeguard: "How often is a lifeguard required to save someone from drowning?" The ludicrous extension of this logic would posit that there should be no requirement that lifeguards demonstrate an ability to swim. The by now obvious point is that basing public safety job requirements on the perfunctory and mundane makes no sense at all.

Establishing job-related standards should be based on the worst-case scenario and common sense (Equal Employment Opportunity Commission 1992), and therein may lie the rub. Virtually all people with tenure in law enforcement can recite examples of officers who have been locked in physical struggles that resulted in the death of a perpetrator. In a free-flowing and unstructured occupation such as law enforcement, there are endless permutations for the application of physical fitness constructs. Take the example of a foot pursuit, one of the most frequently enumerated physical tasks in law enforcement (Gaines, Fulkenberg, and Gambino 1993). One JTA conducted for the FBI (Davis et al. 1994) noted incidents in which special agents have chased a suspect for distances up to and greater than a mile. The energy costs of performing this type of pursuit have been measured in a small sample and found to be characterized as "arduous work" (Nevola et al. 2004).

Without a doubt, public safety would improve and the fleeing felon and the general public would have a higher respect for police officers if they knew this capability resided in the incumbent workforce. Moreover, one feels compelled to ask a couple of important questions: What is the risk of possessing this physical profile as opposed to the converse of being incapable of running but a few blocks? and, Which profile would more likely produce favorable results or evoke more pride within the ranks?

By definition, "emergency services" implies a capability to transcend the challenges presented by either the victim or the perpetrator. Clearly, an emergency in the civilian sector should not create a second emergency for the responding public safety personnel. Many a public safety official (Shell 2003) has answered the question, "How do you prepare for a situation?" with "We plan and train for the worst-case scenario and hope it doesn't occur. But should it happen, we're prepared." Our reaction should be measured, effective, and overwhelmingly in the direction of ensuring a successful outcome. Regrettably, for many officers, that simply doesn't happen.

3. The physical capabilities of the criminal element define the activities and response of the law enforcement officer.

Analyses of the criminal population (i.e., the threat) reveal demographics that are largely consistent with the male gender and youthfulness (Davis 2004). Although it's unlikely that criminals routinely engage in a program of physical fitness, it can be argued that youth is a surrogate for fitness. The current practice in establishing physical performance standards is to use incumbents as a normative reference point. This approach ignores the obvious fact that police officers are not confronting, chasing, or arresting each other, and, as a result, norm-referenced standards merely add to the confusion as to where to draw the fitness line. Clearly, the mission (i.e., responding to and dealing with the violent criminal element) should drive the entire standards paradigm.

One of the paramount objectives of training new officers is to provide a prophylactic buffer for personal injury and a superior set of tactics to be used in physical confrontations. In some quarters there is a misguided and mistaken belief that differences in size can be mitigated or ignored through the wizardry of martial arts. The basis for this belief no doubt stems from television and movies in which the slight of frame can disarm and subdue individuals up to twice their size. This fantasy has given rise to the notion that all one need do is raise the skill level of the small individual and the playing field will somehow be leveled. Nothing can be further from the truth. Consider how in every martial arts sport (boxing, judo, wrestling) "fairness" imposes the assignment of body weight classes to the competition. If this were not the case, the resulting mayhem would provoke public outrage.

Anyone who promulgates the theory that in life-and-death struggles differences in size can be neutralized through skill is simply ill informed. The simple axiom that "big people can and will consistently defeat smaller people" requires little proof.

Unlike in most jobs, failure in law enforcement has potentially egregious consequences. Employers can and must be sensitive to a variety of societal issues in the construction of their workforces; however, fulfilling the essential functions of the job must come first.

In the law enforcement community, we have refined the use of the special weapons and tactics (SWAT) to an art form. SWAT officers are the "9-1-1" response for street officers. But, in most instances, this paradigm is paradoxically upside down. It is the beat officer who is in most need (or peril) for the special fitness training that is devoted to SWAT teams. The patrol officer should be in the vanguard of fitness. The SWAT team has numerical superiority, better weapons, and intelligence. The patrol officer

is frequently thrust into uncertain circumstances, not of his or her making, and must establish control until reinforcements arrive. Skills make and mark the difference between success (and survival) and failure, injury, or worse.

4. The physically fit officer has a greater probability of success than his or her less fit counterpart (i.e., more is better).

Part of the process of establishing the necessity for physical performance standards is demonstrating how a desired construct or capacity translates into performance on the job. Further, showing that more of this attribute equates to greater productivity is a basis for ranking or creating a system of merit.

Few would deny the proposition that officers who are more physically fit can run faster and farther and therefore effect the arrest of fleeing perpetrators more successfully than their less fit counterparts. However, proving an obvious statement like this presents a different set of obstacles—and, that a police department should have to go through the effort to validate the obvious is one of the problems with our adversarial legal system.

In the previously cited *Lanning v. SEPTA* case, the defendants embarked on a very expensive and highly persuasive argument to support the premise that more is better. Arrest data for a year were collected and correlated with the specific arresting transit police officers. Because all of the incumbents had fitness data on record, a discriminate function analysis was conducted, demonstrating the clear superiority of the more fit officers. The court was influenced by the power of this information and remarked,

"This Court is not unmindful of the significance of the additional 470 overall arrests and additional 70 Part I arrests that would be obtained if SEPTA's less-fit officers met SEPTA's aerobic capacity standard. For many of the 470 additional arrests, there would be fewer criminals in the SEPTA transit system left to prey on and victimize the rising public. Significant gains in apprehensions and deterrence such as those demonstrated here are to be encouraged and supported by the federal courts. *The Court simply will not condone dilution of readily obtainable physical abilities standards that serve to protect the public safety in order to allow unfit candidates, whether they are male or female, to become SEPTA transit police officers.*" (italics added)

Similarly, in a study of Los Angeles County public safety employees (Cady et al. 1979), the distinct advantages of increasing levels of strength were reflected in the incidence of back injuries. The group with the highest strength had half the injuries of the middle group, who in turn had half the injuries of the group with the lowest strength. This correlation can be extrapolated to the economic costs associated with such injuries, and, given the strength of these associations, one wonders why the public should be willing to assume the economic cost of hiring the weak? Decisions made in the face of such obvious benefits are even more bizarre when one considers that the most physically fit employees are paid the same as the less physically fit.

Ignoring the profound differences that contribute to job-related success does a disservice and injustice to those who possess more of those valuable attributes and abilities. In fact, more is indeed better in law enforcement.

5. Physical fitness is an attribute that, if not maintained, degrades over time.

The General Adaptation Syndrome (GAS) (Selye 1985) is an immutable law of nature that can be paraphrased thusly: What you do not use, you lose. And, what you use becomes stronger. In designing hiring standards, employment opportunity law is driven by the "minimalist" model that attempts to answer the question, "What is the minimum necessary requirement to perform the job?" Although this assumptive

model works well in some applications, it does not work in the emergency services physical capabilities realm. Figure A.1 is illustrative of this point.

Comparing on-entry starting points for KSAs (knowledge, skills, and abilities) and fitness levels reveals a marked dichotomy. At the time of hire, the typical prospective employee is devoid of job-related skills and knowledge; in fact, the person is at the nadir of his or her professional experience. The upward trend line for the KSA data indicates the rapid assimilation of job-related KSAs as a consequence of training and on-the-job experience. The slope of the curve in the first few years can certainly be subject to discussion and manipulation. However, this is not the case with regard to fitness because, without an ongoing and significant effort directed at offsetting the effects of aging, the physical fitness trend line never moves upward. Stamford and colleagues (1978) documented that within one year of appointment and training, a group of police officers was less fit than on appointment to the academy.

The expected rate of loss of aerobic capacity in the general population is 1 percent per year (Kasch et al. 1990). Increases in stored adipose tissue (fat) was evidenced in percent of body fat and reflected in total body weight can be striking (Davis and Starck 1980). It is not unusual to see public safety employees add more than 50 pounds (22.6 kg) to their frames before reaching retirement. Not surprisingly, fitness degradation places a significant toll on joint loading, and

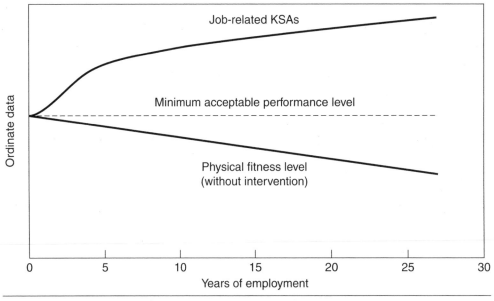

Figure A.1 Contrasts in knowledge and fitness over employment history.

it also significantly interferes with ambulatory activities such as running and walking or standing a fixed post. In addition, numerous metabolic diseases are associated with hypokinesis and obesity including hypertension, diabetes, and atherosclerotic disease, to name a few (Blackburn 1980).

That is not to say that there cannot be some temporary improvement to the physical fitness level as a consequence of periodic "rehabilitative" or basic physical training. But such interventions are short-lived. Once a person has graduated, the normal course of action is to retreat to the previous sedentary state.

Figure A.2*a* shows a person with a physical fitness level that starts well above the minimum acceptable performance level. That fitness level degrades linearly over time without intervention, depicting an officer who is physically "coasting" while employed. In this model, the person's "buffer" above the minimum acceptable performance level is steadily lost over time.

In the scenario shown in figure A.2*b*, the officer comes to the job with a fitness level that exceeds the minimum and, while on the job, participates in a regular program of physical activity. The decay in fitness across the span of employment is barely discernable. This well-known phenomenon has been reported in numerous longitudinal studies of the general population of adults who exercise on a regular basis (Kasch et al. 1990).

6. There are no prohibitions against applicants or incumbents improving their physical fitness commensurate with the requirements of the job.

There is a wide range of scenarios that play out during the hiring process. Of course, the preferred applicant is a physically fit and capable person who presents him- or herself for employment in a highly conditioned state. At the other end of the spectrum are the ill-prepared applicants who believe it is the responsibility of the hiring agency to conduct remedial training; that is, they fail to meet the intake criteria and they file a complaint when and if they perceive (and think others will perceive) that the selection criteria are too rigorous. The end game is obvious: What cannot be accomplished by effort is attempted by legal fiat. In the face of such challenges, we need to remind ourselves that there is absolutely no prohibition against people improving their employment prospects by presenting themselves in a physically fit condition. To the contrary, much as we demand job preparation for other professions, this is exactly what we should expect from law enforcement. Finally, without the imposition of retention standards linked to graduation criteria, much of the physical fitness training that is conducted at the police academy becomes an "exercise" in futility over the longer term.

Unfortunately, once people are hired and trained, "coasting" all too often becomes the normal course of action. Such inaction begs the question, "Why is it

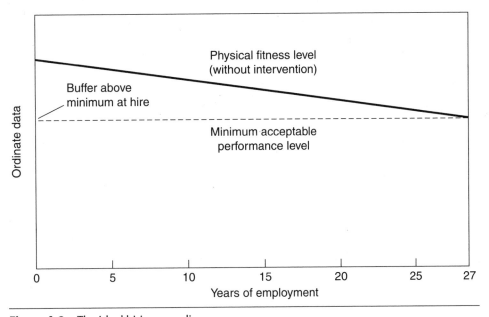

Figure A.2 The ideal hiring paradigm.

the responsibility of the agency to rehabilitate?" Keep in mind that coasting only goes downhill and that physically fit incumbents who work to maintain the basic capacities needed to properly perform the job are clearly those who are the most qualified.

For both cadets and incumbents, maintaining job-related physical fitness is a personal responsibility in the same way that maintaining mental proficiency is a job-related personal responsibility. The use of one's body in law enforcement is a far more likely occurrence than resorting to the use of a firearm. The irony is that in virtually all law enforcement agencies, the use of a firearm requires some form of requalification. Why is there no such requirement for physical fitness?

Virtually all law enforcement agencies have unfit patrol officers who would be better deployed elsewhere. Basing standards on the lowest-performing incumbent harms public safety and perpetuates mediocrity, reminding us of the old adage that "successfully performing one's job does not mean simply showing up for a paycheck." Improving the quality of the workforce is a legitimate governmental function.

7. No physical ability test is ever perfect; however, the use of surrogates in testing for job-related capacities is superior to more abstract measures.

There are several approaches to establishing validity (the process that demonstrates the link between the test for the job and the job itself). The definitions and doctrine for validation grew from the disciplines of educational and cognitive testing. It is not this author's intention to provide a primer on this complex subject, but rather to point out that it can be very complex and that attempting to build a test can be very much like performing brain surgery on yourself. You can do it, but you might not like the way it turns out. Having said that, a number of important ingredients are involved in the development of a good test.

We start with the concept of objectivity. Objectivity is judgment based on observable phenomena, uninfluenced by emotions or personal prejudices. In short, objectivity lets the data speak for themselves. Before anything, we must have objectivity. Objectivity ensures that the observable data (e.g., repetitions, time, or weight) are the same, irrespective of the person administering and interpreting the test.

Next we must have reliability. A reliable test is one that yields the same or compatible results in different clinical experiments or statistical trials.

In the context of target practice, figure A.3 is illustrative of these concepts: validity, objectivity, and reliability. First note in figure A.3a that the group of shots is disorganized. If the objective of the test is to hit the bull's-eye, the objective is not being met, and this test is presumed to be valid in that it appropriately demonstrated the lack of requisite skill in sight alignment and shooting skills.

In the aftermath of an examination of the results of a first trial, the instructor would likely validate these findings with a second trial. If the second trial is confirmatory, we can logically conclude that the student is a poor shot and the test is both reliable and valid.

In the next example, figure A.3b, note that the result of the six-shot trial is a tight cluster just outside the target. The assumption one might logically draw is that the requisite skills were present, but the test failed to produce its intended results; that is, the gun does not shoot straight.

In the practical application to fitness testing, we might have a tool that yields reliable results, but fails to hit the mark (predicting job performance) and

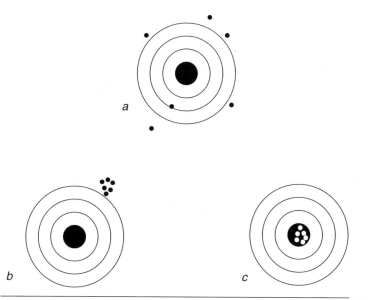

Figure A.3 Target practice: (a) reliability; (b) reliable but not valid; (c) reliable and valid. (Bullets that are clustered but outside of the bull's-eye are reliable. In other words, they're grouped, but not valid since the object is to hit the center of the target.)

therefore lacks validity. In this context, research has failed to demonstrate validity for a number of widely used gym-based tests (Di Vico et al. 2005).

Figure A.3*c* combines the requisite skills and an accurate tool, resulting in meeting the criteria of objectivity, validity, and reliability.

Figure A.4 is illustrative of how a physical ability test purports to meet its objectives. There are a number of approaches to developing job-related selection criteria. Employing a statistical relationship is called criterion-related validity. That is, the test explains significant variance on the job through a statistical association.

Circle A encompasses everything that we can know about the job. The process of divining this information is obtained through a formal job task analysis (JTA). For the sake of this presentation, let's assume that 20 percent of the job includes a requirement for physical activity.

Building a comprehensive taxonomy of the JTA occupational tasks and assigning their relative importance (criticality), frequency, and arduousness is an exhaustive process. Harkening to the important corollary under axiom 2 earlier (i.e., building standards to meet or exceed the worst-case scenario) is the best method of ensuring that a reasonable reserve is available for those urgent and critical situations. Figure A.5 demonstrates this point.

In Figure A.6, the clear circle to the right (the job test) can be interpreted to have a "good and reasonable" overlap with the shaded circle to the left (the physical abilities requirements of the job). The area in common between the circles can be predicted by the test. The greater the overlap, the greater the "explained variance" between the job and the tests,

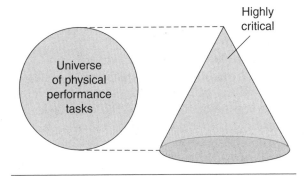

Figure A.5 The hierarchical model: Building around a worst-case scenario.

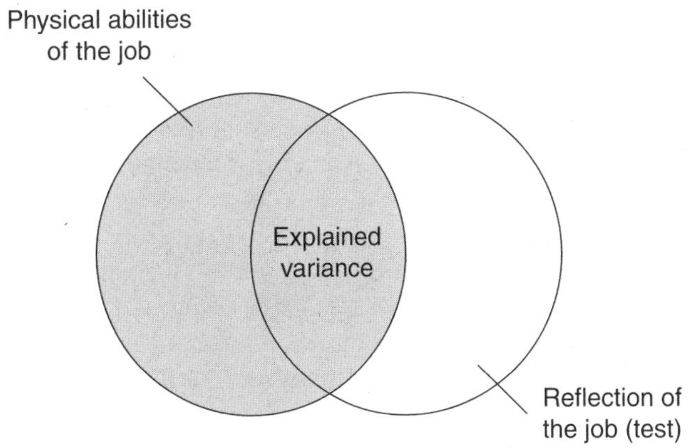

Figure A.6 Explained variance: The greater the overlap between circles, the better.

and, as a result, the greater the power of the test to examine, predict, and explain the test subject's physical abilities to perform the job. The portion that is not covered is the unexplained variance.

Unless we are investigating relatively simple tasks in jobs such as material handling (i.e., in which the job tasks and the job tests can be made to be identical without endangering the test takers and investigators), it is oftentimes necessary to build a surrogate, or facsimile, of the real thing. This is frequently called modeling or simulation. This challenge arises in the public safety sector in which we can never fully replicate with precision all of the permutations of the critical tasks that make up the job, for the obvious reason that the risks of the job preclude us from doing so. For example, we can't allow real bullets or fists to fly. As a result, accurate quantification of the most dangerous and arduous tasks within law enforcement is impossible. To meet this challenge, investigators build simulations that

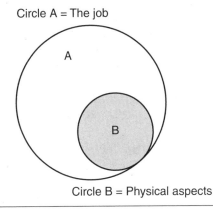

Figure A.4 Job components: The taxonomy of job tasks.

have reasonable fidelity to the actual job tasks or essential functions of the job through the fabrication of a "content-valid approach."

Typically, law enforcement test batteries incorporate a content-valid compilation of "gym-based" tests. The use of body-weight resistive exercises such as push-ups, sit-ups, and distance running is convenient and field expedient, and, as such, they have been frequently and widely used for decades. Although they are not perfect, the validity of such tests can be demonstrated. For example, the earlier-cited research for the NYPD (Di Vico et al. 2004) demonstrated that there was predictive power in push-ups and other similar body-weight tests. This is logical because the muscle does not distinguish between grappling with a suspect or extending the elbow while performing a push-up.

However, because (1) all such content-valid testing has built-in error (i.e., you can't measure with perfect reliability); and (2) no test is perfect or perfectly precise (i.e., virtually nothing approaches 100 percent explained variance); and (3) judges, juries, and attorneys are laypeople, the value and benefits derived from performing such tests are sometimes lost in expert testimony, debate, and litigation. Sadly, objective, valid, and reliable content-valid tests are wrongly declared to be invalid. For this reason, the author resorts to the use of "duck validity," which is proffered and illustrated in figure A.7.

Simply put, duck validity seeks to clarify and corroborate the complex and sometimes contradictory statistical procedures and statements proffered by opposing experts by observing and stating in laypeople's terms what is clearly obvious to everyone.

A recent study (Davis et al. 2006) conducted for the Department of Homeland Security (DHS) and

1. Does it sound like a duck?
2. Does it walk like a duck?
3. Does it look like a duck?

Ergo:

Figure A.7   Duck validity.

now used at the Federal Law Enforcement Training Center (FLETC) is the embodiment of the concept of face, or duck validity. Starting from a belted position in a vehicle, the candidate follows a circuitous course over walls, through windows and culverts, while moving or dragging life-sized mannequins. The clear benefits include the incorporation of on-the-job skills and can be expanded to incumbents with the inclusion of such tasks as cuffing, use of a baton, or even live-fire stations. Interestingly, data indicate that some females perform better on real-world tasks such as the LEOPARD than they do in "pure measures" of physical fitness (Becket and Hodgdon 1987).

8. There is no such thing as a physical ability test that does not have adverse impact.

The framers of the Uniform Guidelines on Employee Selection Procedures, who no doubt were primarily driven by the principles and disciplines of educational and industrial psychology, probably never anticipated the complex issues associated with disparate impact when applying the guidelines to physical ability testing in the public safety arena. The problem arises from the assumption of equal ability and equal results, which is a virtual impossibility in the realm of fitness-related performance for a number of physiological factors, not the least of which results from the significant differences, on average, between the sexes. Figures A.8 and A.9 are illustrative of this point.

The discrepancies between males and females on gross measures of aerobic capacity and strength (lifting capacity) indicated in figures A.8 and A.9 are emblematic of the problem of mitigating adverse impact in hiring standards. Adverse impact is defined by the "four-fifths rule"; that is, if less than four-fifths of a protected class (e.g., women or other defined minorities) are able to pass the test, it has an adverse impact. A test need not be validated unless it is demonstrated to have adverse impact. This is, in and of itself, a Catch-22 because a test, by definition, must have some power of discrimination. In other words, why give a screening test if it never screens anyone out? So, by this definition, we must conclude that any test worth giving will have a disparate impact on someone.

The data displayed in figures A.8 and A.9 were collected from incoming recruits for the U.S. Army (Vogel, Wright, and Patton 1980). The vertical lines, identified as L, M, and H, are representative of the job assignments known as an MOS (military occupational specialty) that correspond to low, medium, and high aerobic and strength requirements. For

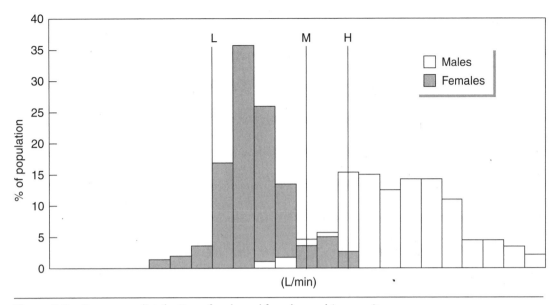

Figure A.8 Frequency distribution of male and female aerobic capacity.

Data from J. Vogel, 1979, *Development of gender-neutral job standards* (Notick, MA: U.S. Army Research Institute of Environmental Medicine).

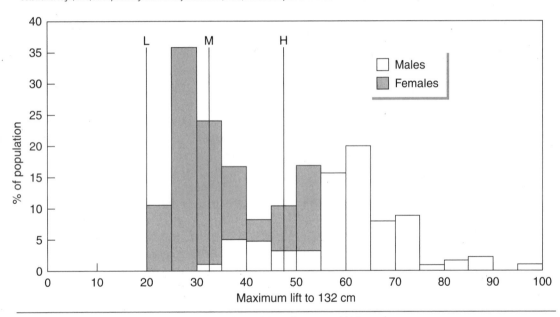

Figure A.9 Frequency distribution of male and female maximal lifting capacity.

Data from J. Vogel, 1979, *Development of gender-neutral job standards* (Notick, MA: U.S. Army Research Institute of Environmental Medicine).

example, the infantry job classification is high in both strength and aerobic capacity. Based on these data, it is clear that (1) the average male recruit is stronger than the average female recruit, (2) many women are of equal strength to men, (3) some women are stronger than some men, and (4) the likelihood of a new female recruit possessing the requisite level of fitness to satisfactorily perform the infantry MOS is about 2.5 percent of the unremediated female recruit population.

9. The fact that a test has an adverse impact does not make it illegal.

Returning to the definition of adverse impact, a test is proclaimed to have an adverse impact if a protected class has a pass rate that is less than four-fifths of the majority. Declaring a test to have an adverse impact is the prelude to validation, or demonstrating that there is a compelling necessity for the constructs, dimensions, capacities, or abilities that are essential to the job.

When a physical ability test does what it is intended to do (provide a ranking of requisite abilities and screen from employment those who do not possess the minimum necessary levels of fitness), sometimes there can be a reaction that the test must somehow be flawed. Unfortunately, some people believe that physical ability tests that do not have an adverse impact on women can be developed within law enforcement. To accomplish this objective, the norm-referenced pass point is moved to a position at which the test no longer has an adverse impact, often nullifying the validity (and certainly the utility) of the test and leaving the jurisdictional authority with the mistaken impression that the pass score produces competent people. In fact, the obverse is more likely.

As the data in figures A.8 and A.9 demonstrate, it is impossible to design a physical ability test that will not have adverse impact on women. To be sure, in *Lanning v. SEPTA,* the district court and the Third Circuit Court of Appeals agreed that the expectation to run 1.5 miles (2.4 km) in 12 minutes was a job-related requirement. The fact that few of the female applicants arrived prepared to meet that standard did not invalidate the requirement. In support of this point, one need only remember that a young, fleeing suspect does not slow down his or her flight to match the speed of the pursuing officer. Very simply, and as a general rule, law enforcement job requirements and combat with a fleeing and resisting felon are not "adjusted" to allow everyone to participate.

The reader will be interested to know that, as an activity, running has the lowest potential for adverse impact, because it is one's own body weight that provides the resistance. Further, there is significant research demonstrating significant improvements in fitness for both genders as a consequence of activity-specific training. With considerable interest in the benefits derived from improvements in physical work capacity, the U.S. military has conducted numerous studies of the female response to training. In fact, with training, improvements in strength and cardiorespiratory fitness can reach as much as 16 percent.

10. There can be no physical fitness without a physical fitness program.

Most of us are aware of the numerous studies that have shown time and again that our increasingly sedentary society is increasingly unfit. Police officers are no exception. Consider these consequences: Cardiovascular disease and back injuries are the leading causes of premature retirement. Lack of training and physical fitness exacerbates the injuries suffered and costs resulting from assaults on officers that, according to the FBI's "Law Enforcement Officers Killed and Assaulted" (2004), averaged more than 56,000 per year over the past 10 years. We lose many more of our officers to poor lifestyle decisions and preventable disease than to gunfire. Lack of physical activity has overtaken smoking as a major risk factor for heart disease. Obesity has a number of profound consequences. From a metabolic perspective, the risks of hypertension and diabetes are a multiple of the threat to individuals of normal weight. The epidemic of obesity is now a regular staple in all the health, medical, and general news media across the nation. Millions of Americans and youth are routinely failing fitness tests. It is estimated that regular soda drinkers will gain an additional 17 pounds (7.7 kg) over eight years, and that 90 percent of consumers drink carbonated beverages regularly.

If ever there was a profession that society expects to be on the cutting edge of fitness, it is our public safety personnel. Yet, a cursory glance around the squad room reveals some significant offenders. This condition is not going to change without leadership. Our society has already validated that, left to their own initiative, most people will hunker down, eat copious amounts of unhealthy food, and not exercise.

Changing attitudes and behaviors is one of the most difficult things that a law enforcement leader can do. Change will not come easily, or by itself; it must be programmed. By definition, a program is a system of services, opportunities, or projects designed to meet a need. You cannot divine that your department is fit to perform the essential functions of a law enforcement agency without some criteria against which to measure job-related constructs. The initiative for making a change is within the purview of leadership and management. Failure to include the rank and file is a formula for failure.

Program implementation and management is a well-researched area of study and beyond the scope of this discussion. However, the following maxims continually rise to the top:

1. Aggressively recruit and hire the best people you can find. They are out there.

2. Then, develop a long-term strategy for success. Your officers will not and did not become unfit overnight. Likewise, they did not and will not

instantly become fit. Consider age-based variations and "grandfathering." Set realistic goals. Expectations to lose weight at a rate of more than 2 pounds (0.9 kg) per week are virtually impossible to obtain and sustain.

3. Lead by example. "We're all in this together" is a great way to start the ball rolling. The chief doesn't have to be the most fit officer in the department. But, a successful program certainly requires that senior management be very much in evidence and that they be good leaders and examples for the rest of the department.

4. Encourage and reward success. Incentives and positive reinforcement go a long way. Until your pay scale rewards fitness rather than mediocrity, it is essential to find ways to motivate and reward excellence.

5. Treat fitness as a benefit. Sustaining a fitness program is a marathon, not a sprint. Remember that if you're not sweating sufficiently to require a shower, you're not working out! There's simply no magic pill or substitute for the expenditure of effort to improve or maintain fitness.

6. Make your skills training and physical fitness programs relevant. Abstract or obtuse measures that purport to demonstrate physical readiness do not generate "traction" with most employees. Showing a

relationship between the measurement instrument and the job is one of the best methods of garnering participation and increasing the potential for officer survival.

7. Seek out and retain competent consultants. With all due respect to the invaluable services the health club industry provides, such individuals are simply not qualified to develop and implement content-valid, JTA-based law enforcement fitness programs that are objective, valid, and reliable. Developing and implementing a law enforcement program that will meet the continually growing and changing requirements of emergencies and disasters takes involved leadership and the help of a skilled, competent, and experienced resource.

The purpose of this paper has been to present 10 Employment Axioms that are relevant to today's law enforcement executive. Failure to recognize these factors in the design of recruiting, hiring, and retention programs will lessen police and public safety and negatively affect the law enforcement mission and results.

In his paraphrase of the words originally penned by St. Paul, Somerset Maugham said, "The race is not always to the swift, nor the battle to the strong—but that's the way to bet."

Paul O. Davis, PhD, © 2007

Appendix **B**

Functional Testing in Selection, Placement, and Return-to-Work

Paul C. Di Vico, EdD • Health Metrics, Inc./ARA Human Factors

The graying of the American workforce has induced employers to find more creative ways to maximize productivity, reduce health care and workers' compensation costs, improve morale, and still comply with federal and state employment law guidance. Functional testing for physically demanding jobs (e.g., structural and wildland firefighters, police officers, manual laborers, production employees, public works personnel) enables the employer to better match the physical capabilities of the candidate or incumbent employee to the physical demands of the work. More specifically, it permits the employer to do the following:

- Ensure the suitability of selection or placement based on functional (physical) ability relative to work demands.
- Baseline preinjury ability to establish a legal basis for medical apportionment.
- Achieve a more accurate determination of pre-injury status in return-to-work applications.
- Enhance assessment of treatment outcomes and establish more cost-effective case management.
- Create an effective transition for workers returning to a job postinjury and minimize the risk of reinjury.
- Ensure the necessary documentation for workers' compensation, disability, and health insurance applications.

It should be noted that functional testing has taken place in two historically different environments: the preemployment environment and the post-offer-of-employment environment. An in-depth description of the differences between the two is well beyond the scope of this appendix and better left to those with formal legal training. The primary difference relates to the questions the employer can ask the candidate prior to administering the assessment, particularly with regard to current or past medical history.

The purpose of this appendix is to present an overview of the concept of functional testing as applied to physically demanding occupations. Thus, I limit the narrative to matters pertaining to what the functional tests are, what they are designed to do, and how they may be integrated into a systems approach to hiring and retention decisions specifically oriented toward maximizing employee job performance and minimizing occupational injury risk. Figure B.1 illustrates the relationship between the essential functions of the job and the "fit" of a given worker's physical capabilities in relationship to functional testing. There are four key components to the process:

1. Job task analysis: Determining the essential functions.
2. Development and validation of the functional testing protocol, the measurement tool for analyzing the fit between a worker and the job.
3. Determination of the performance standard(s) associated with the test protocols: What are the performance criteria indicating satisfactory or unsatisfactory job-specific work capacity?
4. Program policy and action steps
   a. What to do when the fit is good
   b. What to do when the fit is poor

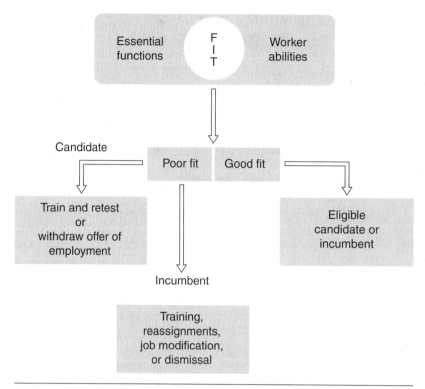

Figure B.1   Functional testing determines the "fit."

The job-relatedness of both the test model and the established performance standard is critical to both the effectiveness of the testing program and its scientific validity and thus its legal defensibility. The Age Discrimination in Employment Act, the Americans with Disabilities Act, Title VII of the Civil Rights Act, the Rehabilitation Act, and the Uniform Guidelines on Employee Selection Procedures are all relevant considerations when developing a job-related test. The four considerations previously noted provide a reasonable operational approach to developing effective, valid, and defensible tests.

Job Task Analysis: Determining the Essential Functions

Three primary sources of information will help employers determine the essential job functions and complete the job task analysis:

- Operations supervisors, medical personnel, and human resources personnel can identify the critical work processes and their historical impact on the physical status of the employees.

- Employees performing the work can offer their opinions of the critical functions and "areas of concern" specific to successful management of the work demands.

- Direct work observations permit employers to record data specific to the physical dimensions of the work:
    - Weights, job demand levels
    - Work cycle time
    - Heights, weights, posture, duration
    - Tools used
    - Identified concerns
    - Suggested resolutions

Test Development and Validation

Protocols should be developed to assess employee performance in a content-valid methodology. These should then be **prototyped** and reviewed with the employer's designated subject matter experts (SMEs). The SMEs are typically experienced senior operators familiar with the tasks and demands of each job's or job cluster's essential functions. The test protocols should be piloted to ensure that they reflect the job's essential functions, are logistically practicable, can be administered in a manner that results in both objective and reliable data, and above all, can be administered safely.

Performance Standards

Employers should be aware of the requirement that any employment test that has or is likely to have an adverse impact on a protected group (e.g., females, older people) must be supported by documentation that demonstrates the validity of the test model itself (i.e., it must reflect essential and nonmodifiable essential functions of the job). There must also be documentation that demonstrates that meeting the performance standard indicates a reasonable likelihood that the candidate or incumbent employee can manage the physical requirements of the job (i.e., the cutoff score must be appropriate to the job demands). Simply put, supporting documentation must support the premise that the test is valid as both a test model and as the standard required for the job.

Once a consensus is achieved among the SMEs and the test design experts, interim criterion standards should be established through appropriate methods. As much as possible, these should be based on real-time measurements of the work demands (i.e., cycle times; positional tolerance requirements; loads and couplings to be managed; distances moved; necessary movements such as pushing, pulling, lifting, and carrying). The test should be cross validated; the proposed criterion standards should be assessed as they relate to performance when the test is administered to a subset of the populations assigned to the job .

From this point, the employer can explore the ramifications of the criterion standards from the perspectives of adverse impact, necessary employee census, sensitivity, and specificity of the test's predictive capabilities. Then, the cutoff score(s) that best match the overall requirements of both employment law guidance and the productivity requirements of the employer can be determined.

Program Policy and Action

When a candidate or an incumbent successfully meets the requirements of the test, the employer may consider the candidate eligible for employment from a physical performance perspective and move the candidate forward into other appropriate phases of the hiring process (e.g., background check, cognitive testing, drug screening, interview). Administration of the hiring process varies according to the needs of each employer. This author suggests that the expense of each component of the process be considered before the order of administration is determined. However, if the physical nature of the work is known to be challenging, cost efficiency might be well served if the test is administered as early in the hiring process as possible.

If a candidate does not pass the test, the employer has two choices: to deny employment or to provide information specific to the identified performance deficiencies and offer a retest after appropriate interventions are executed (e.g., preplacement conditioning), which could be at the candidate's expense.

For incumbents, the outcomes are a bit more complex. As noted in figure B.1, when a gap exists between ability and work demands, the employer must consider reassignment to a different position, referral for work conditioning, job modification if reasonably practical, or, in the absence of viable alternatives, dismissal. Of course, any actions along these lines are subject to the strong influence of labor and management agreements and associated union contracts.

Summary

The role of functional testing is critical to the development and maintenance of an able-bodied workforce in physically demanding occupations. However, a quantifiable and scientifically valid approach coupled with a prudent policy of application and impact-adjusted outcomes is absolutely necessary if the program is to find a meaningful place within a given business entity or municipal or government agency.

Appendix C

A Chief's Personal Narrative

Douglas MacDonald, Chief • Los Alamos County Fire

My fire service career started in 1971. At that time, our department actively engaged in the Royal Canadian Air Force physical fitness program. This program included push-ups, sit-ups, pull-ups, and a 1.5-mile (2.4 km) run. Nothing in the program was related to essential tasks. From there we progressed to the YMCA program, the Cooper testing program, and then to NFPA's fitness recommendations. I recall, as both a union firefighter and a chief fire officer, asking myself why there wasn't some job-related physical agility test for firefighters. This is not to state that running, push-ups, and sit-ups aren't a valid and necessary component of a fitness program; but still, they are only part of the program.

In 1987, as chief of the Casper Fire Department reading fire service publications, I found what the firefighters in our department had been looking for: a complete wellness program that included diet, risk factor identification, exercise, health risk assessment, monitoring, medical screening, and a job-specific essential: a function physical fitness test. The document I refer to is Dr. Paul Davis' criterion task test (CTT). I immediately contacted the union president, and we requested additional information from Dr. Davis. Through the contract negotiations process, the union agreed to this job-specific testing and total health and fitness program on a trial basis.

The trial didn't last long. One year later during labor–management contract negotiations, the union requested moving from the Cooper testing format to the Davis CTT and health fitness program. As with any program, some people embraced it and some opposed it. One afternoon I received a telephone call from a firefighter's wife concerning the CTT. She was very upset that as chief I was placing her husband in harm's way by making him work out and complete the CTT. She informed me that she intended to bring a lawsuit against me personally if her husband was harmed by this physical fitness program. I attempted to share with her the reasons physical fitness programs are so important to firefighters, the families of firefighters, and the citizens we serve, and that this program would in fact lessen the risk of her husband's vulnerability to injuries on the job. Needless to say, my comments did not persuade her; she ended the conversation with, "I may just sue you even if he doesn't get hurt." One hour later I received another telephone call from another firefighter's wife, informing me that if I didn't get that "fat so-and-so" firefighter her husband was working with into shape, and her husband got hurt because of the so-and-so, she was going to sue me. "And, by the way, Chief, keep up the good work." That was our first exposure to Paul Davis' program.

A few years later, Casper, Wyoming, was a more fit fire department and the CTT national champions for three consecutive years. One of the national champion team members contracted cancer during his fourth year of CTT training. There is no doubt in his mind, nor mine, that because of the fitness program and the combat challenge, he recovered relatively quickly from his bout with that deadly disease.

Through a national search in 1992, I became fire chief of the Los Alamos, New Mexico, County Fire Department. This department provides fire services through a contract with the Department of Energy (DOE). The contract required a physical fitness program for the firefighters, and although the department had held this contract for two years, there was no validated or established physical fitness program in place. Once again I called on Paul Davis to help meet the mandates of the DOE contract. Again, without the union's assistance with respect to the implementation of the wellness and fitness testing program, the process would have been far more difficult to accomplish. Dr. Davis conducted a site-validated physical agility test and provided us with medical standards for firefighters and the on-target fitness program. A three-year phased approach was used to reach the ultimate goal of completing the CTT in 6 minutes, 59 seconds or less, and this was written into the union contract as a condition of employment for combat firefighters.

Although there was support for the CTT among union leadership, there was some contention in the ranks of the union membership. Multiple challenges were brought on many fronts, two of which eventually reached a final determination by the Equal Employment Opportunity Commission, and one of which was decided in the federal Tenth Circuit Court of Appeals. Additional information on those two matters is provided at the end of this appendix. The challenges to the CTT that were brought to the attention of the DOE and the New Mexico congressional delegation were eventually resolved through administrative investigation, including interrogatories and correspondence.

The CTT is an objective, nondiscriminatory way of assessing whether applicants have the skills and abilities that match a position's requirements. The fitness program is life-changing: It creates an awareness of one's lifestyle with respect to diet, activity, stress levels, and exercise habits, and it provides the education tools to help firefighters and their families

achieve better health and fitness and, ultimately, firefighter safety. Our department's fitness levels have been enhanced over the years as a result of the program (see figure C.1).

One of the oldest career combat firefighters in the nation, a member of our department, successfully passed the CTT at age 67 with a time of 6 minutes, 40 seconds. His comment was, "Chief, you know that I can do it faster than that, but I just walked through it." I can't emphasize enough the value of a comprehensive job-related, site-validated fitness program for firefighters, their families, and the communities they serve, and the value of union support in implementing a successful program.

EEOC Determination and Tenth Circuit Holding

Two complaints were filed with the EEOC, one by the union on the basis of disparate impact based on age and gender and one by a woman on the basis of gender, small stature, and retaliation for having spoken out against the test (the plaintiff had been terminated for failure to meet the minimum medical fitness for duty requirements for firefighters). The second challenge went all the way to the Tenth Circuit Court of Appeals. In each case, the EEOC investigated the complaint and closed its files with prejudice, finding no violation of the federal nondiscrimination statutes. The Tenth Circuit Court

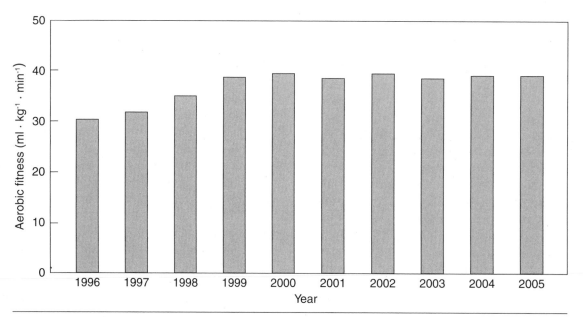

Figure C.1   Aerobic fitness, 1996 to 2005.

also found, in the appeal of the woman's complaint, that there had been no discrimination on the basis of gender, stature, or free speech in speaking out against the test. Following are excerpts from the opinion:

> [Plaintiff] argued below that the CTT had a disparate impact on women and was not job-related. In ruling against her, the district court held she failed to establish a prima facie case. A disparate impact claim involves employment practices that are "fair in form, but discriminatory in operation." Disparate impact claims are typically proven through statistical evidence. Although the record reflects [Plaintiff] herself believed the test was more difficult for smaller persons to perform, she offered no statistics or indeed any evidence demonstrating that the CTT had a significant disparate impact on women. In fact, the only time she took the test she passed it. Moreover, the record indicates that the . . . Department had two other women firefighters and was still requiring essentially the same test. The record contained no evidence showing that a disproportionate number of women were unable to become or remain firefighters with the Department or any fire department as a result of their performance on this test.
>
> [Plaintiff] also asserts she was dismissed in retaliation for speaking out against the test. The district court ruled that [Plaintiff's] speech was not on a matter of public concern and was therefore not constitutionally protected. . . . Speech relating to internal personnel disputes, personal grievances and working conditions ordinarily is not viewed as addressing matters of public concern, while speech that pertains to a public entity's discharge of its governmental responsibilities or questions the integrity of government officials is considered to address matters of public interest. . . . Our review of the record persuades us that [Plaintiff's] statements concerning the CTT were made in her role as an employee rather than as a citizen. Her complaints focused on her own problems with the test. Because her statements were made as an employee rather than as a concerned citizen, they are not entitled to First Amendment protection. Summary judgment was properly granted.

Appendix D

Work Output as a Function of Selectivity in Hiring

Norman D. Henderson, Ph.D • Oberlin College

I. The Basics

Estimating Relative Productivity

Considerable effort has been devoted to estimating the variability in work output (productivity) of workers in many occupations. Hunter, Schmidt, and Judiesch (1990), in summarizing much of this work, found that for medium-complexity jobs, the standard deviation (*SD*) of work output of a group of incumbent workers averaged approximately 25 percent of the group's mean output. This estimate of the standard deviation of work output (SD_{WO}) can be used in conjunction with calculations using a normal distribution (bell curve) to estimate the relative productivity of a single worker or a selected group of workers.

For example, an employee in an organization who is at the 85th percentile in work output (i.e., 85 percent of workers are less productive or valuable) stands approximately 1 *SD* above the employee mean and would thus be expected to be, on average, about 25 percent more productive than the average worker (i.e., 125 percent). In contrast, a worker performing down at the 15th percentile in the group would be 1 *SD* below the mean and would be expected to have a work output of only about 75 percent of the average employee in the group. We also see that the high-performing worker is nearly twice as productive as the worker at the 15th percentile.

Preselected Groups of Employees

Calculations of individual performance can be extended to estimate the average work output of groups of employees. For example, we may want to know how much more productive than average are the best 20 percent of employees in a particular

job. This subgroup would include all employees who range from the 80th percentile, performing about 0.8 *SD* above the group mean, up to the very best employees in the group—above the 99th percentile, who are performing better than 2.3 *SD* above the mean output of all workers on the job. Based on calculations using the normal distribution, the average performance of these top 20 percent of workers is about 1.4 *SD* above the average of all workers. Using the Hunter et al. estimate of *SD* = 25 percent of mean, this group of top workers would be expected to be about 1.4 × 25 percent = 35 percent more productive than the average worker. An even more elite group of workers, such as the top 5 percent, would be even more productive—their mean output is 2.1 *SD* above the group mean, or about 2.1 × 25 percent = 52 percent more productive than the average worker. Two of these top employees can be expected to produce as much as three average employees (2 × 1.52 = 3.04).

Predicted Variation in Work Output of Unselected Job Applicants

In many medium-complexity jobs, incumbents have been through some prehire selection process or have successfully completed a job training program, a probationary period, or a mix of screening procedures. As a result, the variability in output found among incumbent workers is usually smaller than one would find in a group of unselected individuals, such as job applicants. Hunter and colleagues (1990) calculated and then adjusted for this shrinkage effect among incumbents to obtain an estimate of the *SD* in work output expected among a group of unscreened job applicants. For medium-complexity jobs, the SD_{WO} for applicants increased to 32 percent of mean output, compared to the 25 percent for incumbents. Thus, if we were interested in how variable the output of an unselected group of job applicants would be if all were hired, the 25 percent used in the previous computations would be replaced by 32 percent. A person performing at the 85th percentile of a group of workers that had not been prescreened would have a work output approximately 32 percent greater than the group average, and a person at the 15th percentile would have a work output of approximately 68 percent of the group mean. Similarly, the top-performing 5 percent of a group of workers who had not been preselected would be expected to be approximately 2.1 × 32 percent = 67 percent more productive than the average member, rather than the 52 percent estimated for screened incumbents.

Table D.1 summarizes this information about selectivity and work output, both in SD_{WO} units and in percentages of mean output. The table shows how much more productive a selected group is

expected to be, on average, compared to the mean productivity of unscreened workers hired for that job. The percentage of output values in column four are based on an SD_{WO} of 32 percent of mean output, appropriate for applicant populations seeking a variety of medium-complexity jobs. The percentage would be smaller for unskilled jobs and larger for more complex jobs.

From table D.1 we can see that if an employer randomly hired a large number of applicants and then, after observing work performance, let go the least productive 90 percent of the employees, the work output of the surviving 10 percent would be exceptional—20 of them would be as productive as approximately 31 average workers (i.e., 20 employees averaging 156 percent mean work output: 20 × 1.56 = 31.2).

Using a Prehire Selection Procedure

Employers usually don't hire more workers than they need and then cull a large number of the least pro-

Table D.1 Expected Increase in Work Output (or Value of Work Performance) as a Function of Selectivity

Selectivity (top P%)	Cutoff point percentile	Increase in mean work output of the selected group over the mean of randomly hired workers, in SD units	Percentage of greater work output of the selected group, relative to the mean of randomly hired workers*
1%	99th	2.67	85%
5%	95th	2.06	66%
10%	90th	1.76	56%
20%	80th	1.40	45%
30%	70th	1.16	37%
40%	60th	0.97	31%
50%	50th	0.80	26%
60%	40th	0.64	20%
70%	30th	0.50	16%
80%	20th	0.35	11%
90%	10th	0.20	6.4%
95%	5th	0.11	3.5%
100% (None)	0	0.00	0%

*Based on the Hunter et al. (1990) finding that the SD of work output is 32% of the mean output of applicants in medium-complexity jobs.

Adapted from J.E. Hunter, F.L. Schmidt, and M.K. Judiesch, 1990, "Individual differences in output variability as a function of job complexity," *Journal of Applied Psychology* 75: 28-42.

ductive ones. Instead, they use a selection (screening) procedure that is a valid predictor of job performance and hire workers based on scores derived from the procedure. If the selection procedure were a perfect predictor of job performance, table D.1 could be used to determine the productivity gains accrued from selecting the top-scoring P % of applicants from the list. For example, if there are 50 jobs and 1,000 applicants, the employer could simply select the top 50 (5 percent). Alternatively, the employer might set a screening cutoff point that passes 800 (80 percent) of the applicants and then choose by lottery the 50 who will be hired from this group. From table D.1 we can see that the second strategy produces a less valuable workforce (expected gain of 11 percent) than the first strategy of choosing the top 5 percent of applicants (expected gain of 66 percent).

But selection devices are far from perfect. Typically, corrected validity coefficients of screening tests range from .4 to .7 for most jobs.[1] Dealing with imperfect predictors is not difficult—we simply multiply the work output percentages (or SD_{WO}) shown in table D.1 by the corrected validity coefficient of the screening procedure. For example, if the corrected validity coefficient for the screening procedure is .6, the expected productivity gains from choosing the top-scoring 30 percent of the applicants would be .6 × 37 percent ≐ 22.2 percent.

We can apply this procedure to the top 5 percent versus the top 80 percent illustration given earlier, assuming a screening test with a validity of .6. Rather than productivity gains of 66 percent and 11 percent, which would have been realized with a perfect predictor, gains would be .6 of that, or approximately 40 percent and 7 percent, using our less-than-perfect predictor. Note that even with a fallible predictor, the advantage of choosing the top 5 percent versus selecting randomly from the top 80 percent is substantial and remains so, even when validity coefficients are low. It is only when predictive validity approaches zero that the gains from being more selective are lost.

With a large workforce, even small differences in where the top-down cutoff point is set can make a difference in how valuable a selection device can be. Passing 60 percent of the applicants using a screening procedure with a validity of .5 will select a better group of employees than would passing 80 percent of applicants. The first provides an 8 percent productivity gain over no screening, and the second provides a 4.5 percent gain in worker output.

Utility of a Screening Procedure: The Product of Selectivity and Validity

It should be evident from the preceding examples that the usefulness, or utility, of a screening procedure—the gain in work output realized by systematic selection—is a joint function of the validity of the screening procedure and the cutoff point percentile (selectivity) used.

Higher test validity ⟶ Higher utility

Higher cutoff point ⟶ Higher utility
(greater selectivity)

And,

$$\text{Utility of a screening procedure} = \text{Validity} \times \text{Selectivity},$$

where validity is expressed as a corrected correlation coefficient, selectivity is expressed as the mean z-score of the selected group,[2] and utility is expressed as the average gain in work output obtained from the selected group. Because utility is a product of validity and selectivity components, utility drops to zero when either component is zero. Even a test with perfect validity ($r_{xy} = 1.0$) has no utility when all applicants are selected. Conversely, if the validity of the screening device is zero, being highly selective serves no utility function.

Top-Down Hiring Versus a Passing Cutoff Point

Because a top-down hiring procedure automatically keeps the cutoff point percentile at its highest possible value for the number of hires, the utility of the screening procedure is at its highest using top-down selection. If an employer hires 20 percent of the applicants top down, the default cutoff point is the 80th percentile. The increased work output of this group can be compared to the increased work output of any larger group eligible for hire, based on some pass/fail criterion. Thus, if 90 percent of applicants meet some arbitrarily established passing criterion (not uncommon in civil service exams) and the test has a validity of .6, selection gains are 21 percent (i.e., .6 × 35 percent) using top-down selection and 3 percent (i.e., .6 × 5 percent) when hiring is done randomly from the group of 90 percent of applicants who "passed" the test. From a practical standpoint, for every five hires using top-down selection, the employer would get the equivalent of the services of one "free" employee (i.e., $5 \times 1.21 = 6.05$), whereas the productivity gain from hiring five employees chosen from the large pool of 90 percent of passing applicants is trivial ($5 \times 1.03 = 5.15$).

Limitations and Caveats

There are some caveats and limitations to these predictions. First, these calculations are based on a normal distribution, whereas the distribution of performance will often deviate from normality. In those cases, the actual percentage gains will differ somewhat from those listed in table D.1, although the general pattern will be the same. Second, there are often some "false negatives" among the lowest scorers—people whose low scores on the selection procedure are due to something other than low ability. For example, a certain percentage of applicants who appear for physical ability screening show up ill, injured, unprepared, or in other temporary conditions that lead to very low scores. Other applicants are disqualified for reasons unrelated to physical ability such as failure to follow instructions or "over the yellow line" violations. For this reason, the small gains shown in the table for very low cutoff points are probably overstated. Finally, the relationship between screening scores and job performance may not be linear throughout the whole range of scores, although most research suggests that linearity is usually a good approximation, at least when based on incumbent samples. In the next section I argue that, in unselected applicant populations, the linear compensatory model must necessarily break down at the low end of the score distribution.

II. More Complicated Situations

Multiple Predictors in Selection

Many job screening procedures involve assessing abilities in two or more distinct domains, and performance in each is considered in selection decisions. A physical ability assessment is often combined with a measure of job-relevant cognitive abilities in screening for police and fire service, for example. The utility of a multicomponent screening procedure depends in part on how the individual components are combined to make hiring decisions. The procedure that retains the most information involves (1) converting scores from each component into a standardized (z-score) scale[2]; (2) weighting each component based on weights established in the job analysis or from

coefficients obtained from an empirical validation study; and (3) computing the weighted sum, which is often converted to a scale with a predetermined mean and SD. This procedure, based on a linear compensatory model, provides a continuous scale of composite scores containing all the relative performance information from each scale.

Alternatively, the multiple-cutoff method can be used for screening when there are two or more predictors. In this method, pass/fail cutoff scores are established for each component, and only candidates who pass all components become eligible for hire. A multiple-hurdle approach has the advantage of eliminating candidates with critical deficiencies in one ability domain who might otherwise be hired as a result of high scores on the remaining parts of the selection procedure. Often, however, it is difficult to establish either a clear rationale or a sound methodology to determine where a cutoff point should be established, especially when a predictor has a more or less continuous positive relationship with job performance (see Guion 1998, pp. 526-531, for a discussion of sitting cutoff points). Lord (1962, 1963) has also shown that the multiple-cutoff method is technically appropriate only when predictors are perfectly reliable.

Information Loss

A disadvantage of any procedure that uses pass/fail (P/F) cutoff scores is that information is lost, because the continuous variation in applicant scores is reduced to two categories or scores—pass or fail (1 or 0). The relative information loss is expressed by the correlation between the continuous scores and the 1-0 P/F score (r_{cd}). For example, if 50 percent of the candidates pass a dichotomized P/F measure, the correlation between this 1-0 score and the underlying continuous scale is .80. As the cutoff point is moved away from the 50/50 midpoint of the distribution, the correlation between the dichotomized P/F scale and the continuous scale decreases (see table D.2). The effect is small when the cutoff point is near the middle of the distribution, but r_{cd} drops rapidly for more extreme cutoff points.

The loss of information resulting from dichotomization reduces the validity coefficients of pass/fail scales in a straightforward manner—the original validity coefficient r_{xy} is reduced to ($r_{xy} \times r_{cd}$). Thus, at best, the validity coefficient of a P/F scale is only 80 percent of the validity of its underlying continuous scale, and it drops to less than half the original

Table D.2   Loss of Information Reduces Validity Coefficient

Dichotomization cutoff point (% passing or % failing)	50%	60%	70%	80%	90%	95%	98%
Correlation between the dichotomized P/F scale and the original continuous scale (r_{cd})	.80	.79	.76	.70	.58	.47	.35

continuously scored validity coefficient when the pass rate exceeds 95 percent.

Mixed Continuous Scale + P/F Cutoff Selection Strategies

Although it is difficult to defend multiple-cutoff methods when all screening components are scored dichotomously, mixed strategies, involving a continuous scale on one predictor and a P/F cutoff on a second predictor, are rather common in job selection. Often a continuously scored test of cognitive ability is paired with a P/F physical abilities test, for example. Applicants who obtain a passing score on the physical test are usually ranked based on their cognitive test score, and hiring proceeds on a top-down basis.

We can compare the utility of this mixed strategy approach to the utility of using a simple linear compensatory model. For this, I have created an example that uses equally weighted physical and cognitive components. I chose equal weighting because this approach appears to be quite robust. Unless validation samples are very large and standardized regression weights differ substantially, unit weighting and computed regression weights work equally well as predictors in new samples (e.g., Cattin 1978; Dawes and Corrigan 1974). In the example, the corrected validity coefficients of physical and cognitive predictors are each assumed to be .5 for a particular job, such as police patrol officer or firefighter. Because cognitive and physical tests tend to be uncorrelated in unselected populations, the multiple correlation obtained from equally weighting the physical and cognitive predictors will be approximately $\sqrt{(.5^2 + .5^2)} \approx .71$. Using these assumptions, I generated joint normal distributions from which I estimated utilities for various selection scenarios. These are summarized in table D.3.

Table D.3 shows the gains in work output (in SD_{WO}) that are expected, compared to random hiring, under various selection strategies and hiring percentages.

If, for example, 60 percent of all applicants pass the P/F component and hiring is then done top down using the continuous scale component until 20 percent of the total applicant pool is hired, the average job performance of the selected group would be expected to be approximately .82 SD above the mean of a randomly hired group. The SD units can be converted to a percentage gain work output, as was done earlier, assuming we have a reasonable estimate of SD_{WO} for the job at hand. If we conclude that Schmidt and colleagues' (1990) 32 percent of mean estimate is appropriate for our job of interest, we would multiply the SD values in table D.3 by 32, as was done in table D.1.

As we move from left to right across the table, we see that, at all P/F cutoff points, the utility of the joint screening device increases as hiring becomes more selective. Moving down the table we see that, although the utility of the composite test tends to be optimized when the two components are approximately equally weighted, utility differences are small, even when weights differ by a 3:1 ratio. Even modestly weighted components are effective in rejecting the lowest scores on that measure, increasing utility slightly.

The argument is sometimes made that setting a P/F cutoff point too high does not allow enough selectivity based on the continuous measure, resulting in lowered test utility. Obviously, this claim is true in extreme cases in which so few pass the P/F component that most of the passing group are subsequently hired, rendering ranking on the second component meaningless. But that scenario is rare in typical hiring situations. The point is more often raised in situations in which the passing rate is closer to 10 or 20 percent. The utility estimates shown in table D.3 indicate that the argument is false when selectivity is in the 10 to 40 percent range, and even failure rates near 50 percent do not substantively reduce the utility of the composite test. In general, cutoff points producing 60 to 75 percent passing rates appear to be quite effective in maximizing the

Table D.3 Expected Performance Increment (in *SD* Units, *Above*) and Relative Weight of Pass/Fail Component (Italics *Below*) in a Mixed Two-Component Selection Strategy [a]

Passing rate for the pass/fail component	PERCENTAGE OF TOTAL APPLICANT POOL SELECTED FOR HIRE			
	40%	30%	20%	10%
95%	.–	.–	.–	.–
	.11	*.10*	*.07*	*.06*
90%	.52	.61	.73	.92
	.18	*.16*	*.12*	*.10*
80%	.55	.65	.76	.97
	.30	*.26*	*.22*	*.17*
70%	.57	.67	.80	.99
	.41	*.35*	*.29*	*.23*
60%	.57	.69	.82	1.02
	.53	*.44*	*.37*	*.30*
50%	.56	.69	.84	1.04
	.70	*.55*	*.45*	*.36*
40%	(.48)	.68	.85	1.05
	1.00	*.70*	*.54*	*.43*
Continuous scores used with both components[b]	.68	.82	.99	1.24

[a] Upper value: Expected increase in average work output of selected group, in *SD* units, relative to randomly selected applicants. Lower value: Proportional weight of P/F component. The table is based on the model in which the two predictors are uncorrelated and each correlates .5 with the criterion measure. At very high passing rates, utility estimates can be severely influenced by a breakdown of the linear compensatory model and are thus not given for the 95% case.

[b] Expected gains obtained by standardizing and summing the continuous scores of the two components. The utility of either single component would be 70% of these values.

utility of the mixed model across a wide range of selection ratios.

In our example we assumed that the two components were equally important for the job, presumably based on the job analysis or a regression analysis in an empirical validation study. As shown in table D.3, setting a high passing rate for the P/F component shifts the two weights substantially away from the equal-importance conclusion of the validity study. Thus, the second consequence of setting high passing rates on a P/F component is that the component weights used in the resulting screening procedure deviate rather substantially from those obtained from the job analysis or regression analysis.

Table D.3 also shows how the utility of the mixed strategy compares with the utility of the linear compensatory model. The bottom row of table D.3 shows the utility of the linear compensatory model, based on the sum of the two component scores. Comparing these utility estimates with those directly above shows that the utility of the mixed P/F + Continuous

Scale falls short of the level obtained by the equal weight linear model at all selection levels.

Lower-Bound Utility When One Predictor Reaches Zero Weighting

In a true linear compensatory model with two predictors, the lower-bound utility would be equal to the utility of the poorer predictor used alone. In our example used to generate table D.3, the two components were created to have equal validity coefficients of .5, so the lower-bound validity would be .50 when one of the predictors has zero weight. The single component validity is thus .50 / .71 = 70 percent of the two component validity. Because utility = validity 3 selectivity, the utility of using a single continuous measure will therefore also be 70 percent of that for the joint predictors. For example, the .68 utility expected when both continuous scores are used to select 40 percent of the applicants would drop to .48 if only one of the measures were used.[3]

Because test utility is not expected to show a severe drop when one predictor is removed from a linear model, some have suggested that a "troublesome" component can be dropped from the selection process with only a modest loss in test utility. In our example, dropping the P/F component altogether would lower our test utility about 30 percent at any hiring ratio. If that P/F component happened to be, for example, an expensive-to-administer physical capabilities assessment that also produced a substantial disparate impact on female applicants, consideration might be given to dropping the measure on the grounds that selection utility loss should not be very large. The difficulty with this approach, apart from the question of whether a 30 percent loss in test utility represents a significant economic loss, is the question of whether the linear compensatory model is ever valid throughout the full range of possible scores for each component of the test.

Although the linear model appears reasonably correct when used in validation studies involving incumbents, some selection has usually occurred on at least one of the abilities being assessed by the predictors. For example, when a linear relationship between a cognitive ability scale and job performance is observed in a group of job incumbents, it is usually based on a sample that does not include people with exceptionally low cognitive or physical abilities. A corrected validity coefficient for the cognitive scale is usually computed by taking into account the range restriction on the cognitive measure found in the study sample, relative to an unselected sample. Although this is an appropriate correction, it nevertheless applies to a sample in which people with exceptionally low physical ability are absent. The converse of this situation is also true—the linear relationship obtained between scores on the physical capabilities assessment and a job performance criterion measure is based on a sample that is devoid of people with very low cognitive ability.

A linear compensatory model that functions well in validation research with incumbents must break down at some point in scoring when applied to an unselected population containing some people with very low levels of relevant abilities. A linear relationship between physical capability and job performance observed among incumbents will not be present in a subgroup of applicants with cognitive abilities substantially below that required to learn and perform critical job tasks. Job performance of all members of this subgroup will be low, regardless of physical ability level. Conversely, a positive

linear relationship between cognitive ability and job performance observed in incumbents should not appear in a subgroup of candidates whose physical capacities fall far below those required for some essential critical job elements.[4]

Therein lies the fallacy in the argument that a valid but "troublesome" measure can be dropped or weighted near zero in selection without a large loss in test utility. The argument is based on the unjustified extension of regression results to scores that fall below the range studied in the validation process. In most of these cases the expected utility of the single remaining measure will be overestimated, and the utility loss resulting from zero or near-zero weighting of a troublesome component will be underestimated.

The breakdown of the linear model also creates a problem for estimating composite test utilities when a P/F component has a passing rate much above 90 percent. In these cases, depending on the proportion of unacceptable candidates in the pool for each component and other factors, the observed test utility can differ from expectations derived from a linear compensatory model.

Concluding Observations

The joint role of test validity and selectivity in choosing the most qualified employees has been understood for more than 60 years, beginning with the work of Taylor and Russell (1939) and later Brogden (1949). This early work has been steadily refined and extended, especially by Schmidt and Hunter (e.g., Hunter, Schmidt, and Judiesch 1990; Schmidt and Hunter 1983, 1998; Schmidt, Hunter, and Pearlman 1982; Schmidt et al. 1979). We have thus accumulated a considerable body of evidence that attests to the enormous economic impact of selective hiring using valid predictors.

Despite this substantial literature that argues persuasively for high selectivity in hiring, many agencies adopt suboptimal selection procedures, sometimes through lack of awareness and sometimes as a result of external pressures, including litigation. Most often the strategy includes reducing selectivity by setting a very low cutoff point on some troublesome component of the screening procedure, such as physical abilities assessment, because of the large gender differences encountered. The effect of the low cutoff point decision is twofold. First, it shrinks the utility of the physical abilities component to near zero with respect to identifying the most physically qualified candidates, even when the validity of the screening procedure is high. Second, regardless of what might have been found in a job analysis or a criterion validation study, the relative contribution of physical scores to selection becomes trivial relative to the other screening components, questioning the validity of the overall screening procedure.

One must also consider the possibility that there are contractual and ethical, or at least "truth in advertising," questions raised when test developers employ extremely low cutoff point strategies. One may fairly ask whether it is appropriate to create a test that serves a substantial public relations function but only a trivial screening function for an ability clearly identified as important for job performance. Any public agency or business should have the option to knowingly sacrifice some test utility to accomplish additional hiring objectives, such as obtaining a diverse workforce. In some cases the long-run advantage of such a strategy may more than compensate for loss in test utility. But these must be conscious and open decisions by community leaders and management, not something quietly accomplished through subtle psychometric manipulation.

Footnotes

[1] Uncorrected validity coefficients are the observed correlation coefficients between scores on the screening test and actual job performance. Observed test criterion correlations are usually smaller than the actual correlations because of attenuating artifacts, such as unreliability of the criterion measure and restriction of range in the employee group. Corrected validity coefficients include adjustments for some of these artifacts. The discussion assumes that validity coefficients have been corrected for attenuating artifacts.

[2] A z-score is a standardized score. It is obtained by subtracting the group mean from an individual's raw score and then dividing (Individual score – Mean score) by the group standard deviation: $z\text{-score} = (X - M) / SD$. Z-scores computed on a group have a mean of zero and a standard deviation of 1. Because all z-transformed measures have an identical SD, variables with widely different scales of measurement can be converted to a common z-score scale for combining into easily weighted test batteries.

Standardized scores are also useful for test utility analysis in that they can indicate the amount of gain work output obtained by selecting a top-scoring subgroup from a larger (applicant) group. When all members of a group are selected, the mean z-score of the selected group is zero. As selection becomes more stringent, with a decreasing percentage of applicants chosen from the top down, the mean z-score of a selected group increases, as shown in table D.1.

[3] Note that .48 is the value shown when the passing rate is 40 percent for the P/F component, and 40 percent of the applicants are hired.

[4] Many criterion measures are based on rating scales where overall performance is defined as the sum of several performance dimensions that are assessed. These rating schemes allow employees to compensate for severe deficiencies in some job dimensions by good performance in others (i.e., a linear compensatory model). Although these summary ratings work well for jobs involving a diversified set of duties of moderate to high importance, they are inappropriate when a job contains one or more critical dimensions (essential functions), in which poor performance is not tolerable. In those cases the incumbent is an unacceptable employee, even when exhibiting high levels of performance in the job areas. A lifeguard who is a weak swimmer is clearly an unsatisfactory employee, even if he receives high ratings on other aspects of the job.

Unfortunately, performance rating scales often contain a mix of critical and less important job dimensions that are simply summed to provide an overall performance rating of each employee. The summation procedure produces an artificial linear compensatory model of job performance that is invalid for incumbents scoring low on critical dimensions. When the job performance criterion measure is based on an invalid linear model, a screening test based on a linear model will not appear to break down, even at extreme levels of low ability.

Appendix E

Biochemical Evaluation of Workplace Stress

Delia Roberts, PhD, FACSM • Selkirk College

Workplace logistics often make it difficult to objectively evaluate workers engaged in complex tasks. However, precise comparisons of cumulative fatigue under various conditions, or even among tasks, can be made using biochemical evaluations. There are a multitude of proteins and signaling molecules, transported throughout the body via the circulatory system, that are reflective of many metabolic processes. At any given moment, a small blood sample taken from a superficial vein can provide the investigator with a snapshot of the physiological status of the worker. Although some of these tests are expensive and time-consuming to analyze, for many parameters less expensive technology is available and can even provide immediate analysis from a single drop of blood, saliva, or urine. Physiological stress and the presence or absence of adaptation can be characterized using biochemical markers of different energy systems—degree of hydration, status of various tissues, or level of inflammation, to name a few. Similarly, levels of stress hormones such as cortisol are important indicators of mental stress with physiological consequence.

An investigation was performed to evaluate the physiological load experienced while engaged in manual reforestation. This occupation had been associated with very high injury rates but had not been previously well characterized. Physiological and biochemical data were collected from 10 healthy male tree planters following 19 (T_1) and 37 days (T_2) of tree planting. Subjects were at rest for baseline measures, collected at approximately 6 a.m., immediately upon rising. Postwork data were collected as soon as possible following the completion of the day's planting, between approximately 4 p.m and 6 p.m. Planters carried loads of seedlings equal to 32 percent of their body mass, and heart rates, recorded every 5 minutes, were between 60 and 75 percent of maximum heart rate for an average of 57 percent of the workday. The physiological consequence of this work was revealed by examining the biochemical parameters.

Blood glucose levels remained at near-hypoglycemic levels throughout the day (see figure E.1). From 21 percent (T_1) to 45 percent (T_2) of the blood glucose samples were less than 4.0 mmol/L. The incidence of hypoglycemia was more pronounced T_2, when 27 percent of the values were less than 3.3 mmol/L. During physical activity the uptake of glucose by muscle is greatly increased. In very long duration events, liver glycogen stores may become depleted and blood glucose levels may fall, in spite of gluconeogenesis. Because of the reliance on glucose as a fuel, hypoglycemia may attenuate the ability of the nervous system to conduct signals and hence contribute to injury susceptibility. Although both attentiveness and motor response are impaired during hypoglycemia, motor response is the more sensitive marker and is slower to recover following restoration of blood glucose levels. Additionally, complex skills such as driving have also been shown to be impaired at blood glucose levels in the range of 4.0 to 3.4 mmol/L. Based on this evidence, it seems likely that the potential for workplace injuries would be increased when blood glucose levels are low, as a result of reduced concentration and reflex function.

Counterregulatory hormone responses are elicited when blood glucose levels fall to approximately 4.0 mmol/L. The expected correlation between cortisol levels and blood glucose levels was observed in the current study ($r = .56$, $p < .01$). The means by which cortisol acts to raise blood glucose levels include blocking the ability of white blood cells (WBC) to

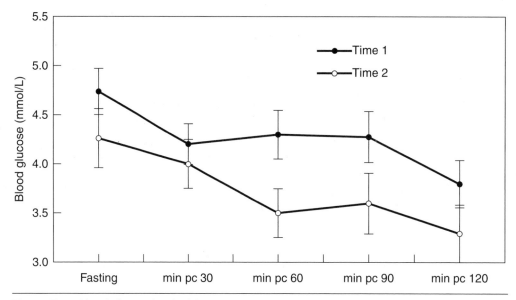

Figure E.1 Blood glucose levels. PC = post consumption in minutes.

take up glucose and increasing the breakdown of proteins for the production of glucose in the liver. Both of these actions would have consequences to planters over the course of a season. The reduced supply of fuel to WBC suggests that the action of the immune system in repairing injuries and fighting infections could be impaired. The immunosuppressive effect of cortisol has been well documented in the clinical literature. In this study a blunting of the leukocytosis (CD4+, CD8+, and CD19+ cells) normally seen after exercise was observed at T_2 coincident to increased cortisol levels (see table E.1), whereas CD3-CD16+CD56+ absolute cell counts were elevated at T_2 rest only. Examination of the WBC differential count suggests that the postexercise leukocytosis at T_1 was primarily due to increased numbers of circulating neutrophils, whereas the elevated resting levels at T_2 were primarily due to increased numbers of lymphocytes.

Muscle wasting in response to increased cortisol levels is also well documented. In the current study, planters lost an average of 1.7 percent of body mass, although daily intake calculated from three-day diet records exceeded 5,000 kilocalories per day, well in excess of the recommended intake for extremely active occupations, including lumberjacks, construction workers, heavy manual diggers, and rickshaw pullers. Skinfold measurements indicated that approximately 75 percent of this loss was due to a loss of fat mass.

Serum creatine kinase (CK) levels increase in association with inflammation and disruption of muscle tissue. Following planting at T_1, CK levels doubled; by T_2, levels remained elevated at rest (see table E.1). These CK values were comparable to those seen in athletes during regular training. Confirmation of the moderate level of CK response is provided by the lack of change in the acute phase proteins C-reactive protein and α1 antitrypsin (see table E.1), which have been shown to increase in response to infection, trauma, or severe exercise.

Other indicators of systemic stress are levels of the catecholamine hormones, epinephrine and norepinephrine. Exercise, low blood sugar, injury, and stress have all been shown to stimulate the release of catecholamines from the adrenal medulla. Increased plasma levels of norepinephrine may also reflect spillover from increased sympathetic nervous system activation. The clearance of these hormones from the plasma following an acute event is extremely rapid; baseline levels are normally restored within minutes. However, during extended periods of stress, elevated resting catecholamine levels are observed. A significant increase in resting norepinephrine levels occurred from T_1 to T_2, whereas epinephrine levels remained more stable (see table E.1). These data confirm the trend toward increased stress at T_2 observed in cortisol, blood glucose, WBC, and CK levels. No correlation was observed in the current study between blood glucose levels and epineph-

Table E.1 Values for Blood Chemistry and Endocrinology

	TIME 1		TIME 2	
	Pre	Post	Pre	Post
Creatine kinase (μl /L)	184±82	409±131* **	265±101*	397±174* **
α1 antitrypsin (g/L)	1.3±0.2	1.3±0.2	1.3±0.3	1.3±0.3
C-reactive protein (mg/L)	<6.0	<6.0	8.6±6.0	9.5±6.7
Cortisol (mmol/L)	428±163	300±147	741±103* ***	300±101
Epinephrine (mmol/L)	0.15±0.15		0.17±0.10	
Norepinephrine (mmol/L)	2.09±1.34		3.09±1.05***	

Data presented as mean ± *SD*.

*Exceeds the normal clinical range.

**Significantly greater than preexercise $p < .05$.

***Significantly greater than time 1 $p < .001$.

rine levels; however, this may have been due to the timing of collection of catecholamine samples (early morning) relative to that of the blood glucose samples (throughout the working day).

Conclusion

Biochemical monitoring of this demanding occupation has revealed evidence for increasing levels of stress over the planting season in the increased CK, cortisol, and norepinephrine levels; immunological data; and loss of body mass with increased number of days of planting. Dietary deficiencies leading to inadequate glucose supply were also indicated. Together these data provide clues to the underlying etiology of the high injury rates observed in this group (Roberts 2002).

Appendix F

Physical Fitness Policy

Paul O. Davis, PhD, and John S. O'Connor, PhD

We regularly receive calls from fire chiefs or training officers who want to implement or update a physical fitness program. They often have a few questions, such as: "We'd like to make this program mandatory, but we have a union that's resisting. What's the best way to deal with this?" or, "If I make this program mandatory and a bunch of my guys just can't cut it, then what happens? Do I have to back off?" or, "Do we need to have physicals every time we do the assessments? That would be too expensive for us."

Let's go, as usual, right to the bottom line: A fire department needs a complete, easy-to-understand policy that is part of the standard operating procedure, written with the involvement and input of affected personnel, and consistently and fairly enforced. If this ideal situation exists, then all of the preceding questions (plus most others) will be answered.

What follows is an outline of what a policy should contain, based on our experience in helping hundreds of fire departments set up their programs plus a collection and synthesis of various policies from around the United States. Keep in mind that these are the key elements; this does not constitute an all-inclusive encyclopedia.

1. **What is expected.** We recommend that the program be mandatory; that is, the members should maintain a minimum level of fitness as a condition of employment. This does not mean that they must exercise on some monitored schedule, only that they demonstrate on a periodic basis they have at least the minimum levels of fitness required to do the job. Fire departments need to take a stand on physical capabilities, just as they do in other aspects of personal performance such as drug use (for which the standard is zero tolerance). Where do you stand on fitness? How much "out of condition" do you tolerate? The answer should be that all members are required to meet the minimum standard of fitness needed to safely perform the job; it should be mandatory. Voluntary programs soon become nonprograms. The time to find out that a person can't get the hose to the floor below the fire is not the day of the fire.

2. **Who is affected.** The policy needs to spell out who is included and who is not. Minimally, everyone who may ever be expected to go out on a call should be covered. Individuals who are expected to wear SCBA should be held to a higher order of performance, because SCBA impose a severe impact on physical performance. Individuals who are not expected to participate in suppression activities cannot be legally held to a performance requirement that is not job related. Ideally, administration and supervisory personnel will set the example and include themselves in the policy. Minimally, they should be visible in their participation, whether held to performance standards or not.

3. **Who is responsible.** The chain of command for ensuring compliance with the fitness policy needs to be outlined. Ultimately, of course, the fire chief is responsible. But the policy needs to spell out the role of the shift commanders, battalion chiefs, fitness coordinator, and medical officer. We recommend that one person be given responsibility for coordinating the program. That way, those who don't meet the standard will know whom to deal with.

4. **The testing schedule.** The policy should describe in detail the tests and how often the members will be tested. Minimally, each member should be tested twice a year, ideally once with a traditional fitness assessment (aerobic fitness, strength test, etc.) and once with a criterion task test such as the

Essential Function Test. The members should know when the tests will be administered and exactly what the standards are, in terms of times, weights, and so on. We believe that corrective action should be taken only for performance on criterion, or job-sample, tasks. Traditional fitness assessments may be viewed as diagnostic, that is, instructive as to why people may not be performing up to par on the job-sample tests.

5. **Incentives and penalties.** This may be the most important part of the policy. All members need to know exactly what the consequences will be of meeting or not meeting the standards. For those who exceed the minimum standards, there should be rewards and recognition. In an ideal world, this would include financial incentives such as a small cash bonus (say $250 to $500 around Christmas) or an additional small merit increase in pay on top of any other regular increases (an amount equal to $250 to $500 over a yearly period). The reality of the fiscal situation for most municipal fire departments dictates that the fire chief may have to reward those who exceed the standards with tokens of recognition (shirts, etc.) rather than cash awards. People who are meeting minimum standards should not be rewarded for doing what they're already being paid to do.

For those who do not meet the required standards, there can be a series of sanctions of gradually increasing severity. However, it's extremely important that these individuals have ample time to come into compliance with the physical performance standards and that there be no penalty after the initial failure, other than that they must begin a monitored fitness program and submit a log of their workout activities. We suggest that consideration be given to a progression of sanctions that would be invoked on a cumulative basis after each assessment at six-month intervals. These could include (1) for initial failure, a requirement to participate in monitored activities and submit a log; (2) after six months still out of compliance, relegation to light duty; (3) after one year, a reduction on the wage scale by one step; (4) after 18 months, suspension for 30 days and an additional wage reduction of one step; and (5) after two years, suspension without pay indefinitely until the person successfully meets the performance standard.

It's important for the members to know that at any time they can reschedule an assessment with the fitness coordinator and, if successful, be immediately reinstated to their former status.

Remember, the law does not require the continued employment of incompetent or unfit individuals. Management has a higher responsibility: mission accomplishment.

6. **Medical clearance.** We do not recommend an annual physician physical or medical clearance for apparently healthy firefighters who are doing the job. However, this assumes that they have passed an NFPA 1582-type medical exam when they joined the department or at the last appropriate scheduled appointment. We do recommend that they complete a health history questionnaire prior to each assessment. This questionnaire is an inexpensive way to identify those who should seek a physician's clearance before the test. (Of course, we have to ask the ironical question: If they can fight a fire, why can't they take a simple fitness test that may be below the demands of an actual incident?)

Some other points: Members should be allowed and encouraged to exercise while on duty. However, all emergency duties, or other tasks as determined by the chain of command, must take priority over the workout time. Also, firefighters returning to regular duty after an injury or light duty for any period should take the physical performance test.

Now for the most important part of this column: When writing and implementing your fitness policy, get the entire department involved. Get all members to buy into the policy up front. And make sure the expectations and penalties are not so stringent that you may have to back off later; this would hurt your credibility and the effectiveness of the whole program.

Select an enthusiastic individual who understands fitness and knows how to relate well to the other members. Let him/her administer the program. Remember to phase in the program gradually and reward successes amply. Be firm but kind with those who initially fail the standards, and guard the confidentiality of everyone's performance.

The keys to success are everyone's involvement up front, good leadership (which includes personal example), and consistent application of a clearly written and understood policy.

Appendix G

Standard Operating Procedures

Don Oliver, Fire Chief • Wilson Fire/Rescue Services, Wilson, North Carolina

1.0 Purpose

The purpose of this policy is to establish guidelines and standards for the implementation of a health maintenance program for all uniformed employees of the Wilson Fire/Rescue Services.

1.1 Wilson Fire/Rescue Services uniformed personnel shall be given at least one (1) hour per shift, for exercising to maintain and/or improve their physical condition. All uniformed employees shall participate in an exercise program tailored to their needs based on their current fitness level. Wilson Fire/Rescue Services personnel will follow a well-balanced exercise program that includes the following major components: (1) warm-up, 5 minutes; (2) aerobic fitness and/or muscular fitness, 20-50 minutes; (3) cool-down, at least 5 minutes. By following these components, all employees should maintain or improve their cardiorespiratory endurance, muscular endurance, flexibility, and body composition.

1.2 The physical ability test shall be reviewed by the committee after each testing cycle as needed. The committee shall consist of the fire chief and volunteers from the membership of the Wilson Fire/Rescue Services. Representatives from the personnel department and city medical staff may serve on this committee when needed.

2.0 Policy

It shall be the policy of the Wilson Fire/Rescue Services that all uniformed employees be physically ready to perform those duties that may be required of them as public safety officers.

2.1 Life Safety Operations personnel will test while off duty.

2.2 It shall be the responsibility of the employee to select their off-duty date and time from the available schedule. An off-duty date and time will be assigned to all employees who fail to meet the posted deadline for selection.

2.3 Employees who fail to keep their appointments for scheduled testing for reasons other than an approved medical leave or personal emergency are subject to disciplinary actions in accordance with the city's disciplinary procedures I-1.

2.4 The Firefighter Combat Challenge will be administered at the Wilson Fire Training Center.

2.5 Employees who are unable to participate in the scheduled testing due to excused absences shall be required to participate on a make-up date, which shall be scheduled within 30 days of the original testing date.

3.0 Definitions

3.1 FITSCAN is a self-contained computer software package designed for processing the results of physical fitness testing and for generating individualized exercise prescriptions. The program incorporates measures of the following:

Aerobic fitness

Body composition

Muscular fitness

CHD risk profile (coronary heart disease)

Flexibility

Pulmonary function

3.2 Satisfactory improvement as used in Section 4.3.8 of this policy shall be defined as the demonstration of adequate measurable advancement toward the goals that were identified in the FITSCAN Customized Individual Training Program. Participants shall be responsible for filling out the duty log sheets that shall be made available by the departmental program coordinator.

3.3 Testing dates shall be defined as those dates that the physical ability test will be administered. These dates shall be scheduled and posted at least thirty (30) days prior to testing.

3.4 Compensation: Employees shall be compensated for off-duty testing by overtime pay or compensatory time accrual. The type of compensation shall be the employee's option.

3.5 The physical ability test shall be defined as the Firefighter Combat Challenge.

3.6 City Medical Staff shall mean the staff of the Safety and Health Division, contractual nursing staff, and/or contractual physician hired by the city Safety and Health Division.

4.0 Procedure/Rule

4.1 Medical Examinations

All uniformed employees of the Wilson Fire/Rescue Services shall be required to take an annual medical examination. Components of the medical examination shall be determined by the City Medical Staff.

4.1.1 Employers are offered the option of participating in the "On Target" assessment scheduled and administered by Wilson Fire/Rescue Services personnel. This testing procedure is an indicator used to predict success on the Firefighter Combat Challenge.

4.1.2 Applicants

Applicants for the Wilson Fire/Rescue Services shall be required to undergo a medical examination prior to employment. In addition, applicants may undergo physical ability testing during the application process. Candidates must meet departmental minimal standards prior to employment. Components of the test will be made available to applicants at the time of application.

4.2 Physical Ability Testing

All uniformed employees of the Wilson Fire/Rescue Services will participate in the annual mandatory physical ability testing during the spring of each year. In order to be excused, an employee must provide a medical waiver to his or her immediate supervisor prior to the testing date. This form is to be completed by the employee's personal physician and returned prior to the test date. A copy of the form shall be forwarded to Safety and Health Division for review and inclusion in the employee's medical file. The employee shall be placed on light duty status until released by his or her personal physician. Once the employee has been released by his or her personal physician, the City Medical Staff shall review all the relevant documentation provided by the employee's personal physician and shall then make a determination about the employee's light duty status. Such determination shall be to (a) return the employee to full duty, (b) extend the period of light duty, or (c) postpone determination pending examination of the employee by a contractual physician of the City Medical Staff.

The Firefighter Combat Challenge will be in conjunction with the spring testing procedure of each year and will be used to select members for any Firefighter Combat Challenge competitive team and for department fitness awards.

4.3 Physical Ability Testing (continued)

4.3.1 Components of the Firefighter Combat Challenge physical ability test shall be related to essential job functions. Those components are as follows:

- A stair climb to the fourth floor (five stories) of the fire tower with a hose pack or high-rise pack not to exceed 55 pounds (25 kg)

- A hose hoist with one section of 2 1/2-inch (6.4 cm) hose not to exceed 45 pounds (20.4 kg)

- A Keiser forcible entry simulator using a 9-pound (4 kg) "dead blow" mall and moving the 150-pound (68 kg) weight a distance of 5 feet (152 cm)

- Advancing a charged 100-foot (30 m), 1 1/2-inch (3.8 cm) hose line a distance of 75 feet (23 m) and flow water

- A victim rescue consisting of moving a 175-pound (79.4 kg) rescue manikin a distance of 100 feet (30 m)

4.3.2 The Firefighter Combat Challenge will award personnel as referenced in the awards policy.

4.3.3 Uniformed employees of the Wilson Fire/Rescue Services who cannot be tested because of preexisting, correctable medical conditions are ineligible for consideration for promotion until cleared for testing by city medical staff, and will be placed in categories identified in section 4.4 of this policy.

4.3.4 Uniformed employees of the Wilson Fire/ Rescue Services who fail to meet departmental minimal standards for reasons other than medical conditions shall be placed on light duty and shall not be permitted to engage in emergency operations. The employee will be placed in an on-duty rehabilitation program administered by the department or city program coordinator for a period not to exceed 90 days. At any time during the 90-day rehabilitation period the employee may retest.

4.3.5 Failure to demonstrate satisfactory improvement or to successfully complete the physical ability test after the 90-day rehabilitation period shall cause the employee's performance to be reviewed by the Chief, who will determine if it is in the best interest of the City to continue the employee's rehabilitation assignment, place the employee on leave without pay, and/or initiate disciplinary action in accordance with the City's disciplinary procedure I-1.

4.4 Physical Categories

Employees under the care of a physician for a medical condition that limits their participation in the physical ability testing process shall provide the appropriate information to the City Medical Staff for review. The employee shall be placed on light duty as described in 4.3 above. The City Medical Staff shall place the employee in the category most appropriate for the employee's condition.

Category I

Problems that, by judgment of the City Medical Staff, are chronically or acutely life threatening or that prevent the employee from performing essential job functions, but that can be corrected by medical treatment.

Employees who are classified in this category shall be limited to light duty and not permitted to engage in emergency operations. When the City Medical Staff identifies an employee who is in this category, they will defer testing, if necessary, notify the Chief immediately by telephone, and follow up with a letter identifying the employee, why the employee was not tested, and the medical recommendations. When the Chief is notified that an employee is placed into Category I, the Chief will review the information and determine if it is in the best interest of the city

to place the employee in a light duty assignment, require the employee to utilize his or her sick leave (if available), or place the employee on leave without pay in accordance with city policy.

The employee may remain on light duty, sick leave, or leave without pay until the employee has met the rehabilitation recommendations made by the City Medical Staff and the employee's personal physician. Once the employee has been released by his or her personal physician, the City Medical Staff shall review all the relevant documentation provided by the employee's personal physician and shall then make a determination about the employee's light duty status. Such determination shall be to (a) return the employee to full duty, (b) extend the period of light duty, or (c) postpone determination pending examination of the employee by a contractual physician of the City Medical Staff.

If the employee fails to make satisfactory progress as determined by the City Medical Staff and the employee's personal physician, and if the employee is in Category II for up to six months, the employee will be placed in Category II.

Category II

Problems that, by judgment of the City Medical Staff, are chronically or acutely life threatening and cannot be corrected by medical treatment and that do not allow the employee to perform essential job functions.

When an employee is placed in Category II, the City Medical Staff will defer testing, notify the Chief immediately by telephone, and follow up with a letter identifying the employee, why the employee was not tested, what problem was identified, and the medical recommendations. When an employee is placed in Category II, the employee may be assigned to light duty, sick leave, or leave without pay not to exceed 90 days until it is determined what course of action will be taken. Courses of action may include the following:

A. File an application for another position within the city that would meet the limitations of the medical condition diagnosed, with no salary decrease to the employee.

B. Show evidence to the satisfaction of the Chief and City Medical Staff that the employee's condition may or may not be permanent, and establish a date for returning to work or for review of the employee's condition.

C. Medical disability retirement.

Appendix H

Essential Functions of Law Enforcement

Paul O. Davis • First Responder Institute

According to the U.S. Bureau of Justice Statistics, there are over 800,000 law enforcement officers in the United States, representing 17,874 state and local agencies and nearly 100 federal agencies. With this diversity, is there a commonality that defines the essential functions of the job? Does every department have to research, design, and implement physical performance standards that are unique to its specific mission? In reality, law enforcement agencies are distinguished by the requirement to "serve and protect"; specifically, to invoke the power of arrest. Noteworthy is the common requirement across the entire spectrum of law enforcement to carry a weapon and apply lethal force.

A cursory inspection of the curriculum for certification as a police officer in any setting—from the FBI Academy or FLETC (Federal Law Enforcement Training Center) to the smallest regional criminal justice academy—reveals a program of instruction (POI) covering the use of firearms, less-than-lethal force, (defensive and offensive) tactics, and physical training. This commonality of critical learning objectives illustrates the point that core competencies already exist. Central to the body of knowledge that has developed since the 1960s is the force continuum, the escalation of force required in securing the compliance of a suspect, violator, or perpetrator. Although there are compelling reasons why the force continuum is more of a set of guidelines than an absolute, infallible, and predictable system of human behavior, the fact remains that the underpinnings of physical ability necessary to effect an arrest are universal. Said another way, it matters little the title, authority, or mission of a law enforcement agency; when a suspect decides to flee the scene or engage an officer in a physical struggle, it is an issue of officer versus subject.

Considerable psychophysiological research has been devoted to "flight or fight" adrenergic responses. But in a contest of wills, in which one party is attempting to effect superiority over the other, it is the physical capabilities, coupled with the application of skill, that will predict the outcome of the contest.

From a number of job task analyses (JTAs) covering an eclectic group of agencies and departments, a clear trend emerges. An examination of the physical nature of the job reveals only a few activities required of police officers that rise to the level of being arduous, critical, and in some cases frequently performed (see figure H.1). Transcending the mundane (e.g., carrying boxes of evidence into court) are those activities involving the pursuit and apprehension of a fleeing or resisting suspect. Incidents in which an officer's very life hangs in the balance elevate the criticality of physical abilities.

The activities associated with the essential function of arrest can be dissected and classified by their muscular and energy requirements. Running is one of the most frequently enumerated physical tasks in law enforcement. The only discussion or debate on this topic is distance and pace. Department chiefs would say that in a one-on-one foot chase, the ideal outcome is when the officer emerges as the victor. There is nothing mysterious about the result of such a contest.

The scientific literature contains numerous studies of the relationship of aerobic and anaerobic power as predictors of the outcome over measured courses. Aerobic power is associated with success in runs exceeding 800 meters, whereas anaerobic power is associated with success in sprints. Running distances longer than a few city blocks will tax the aerobic capacity of an officer; shorter-duration sprints call

217

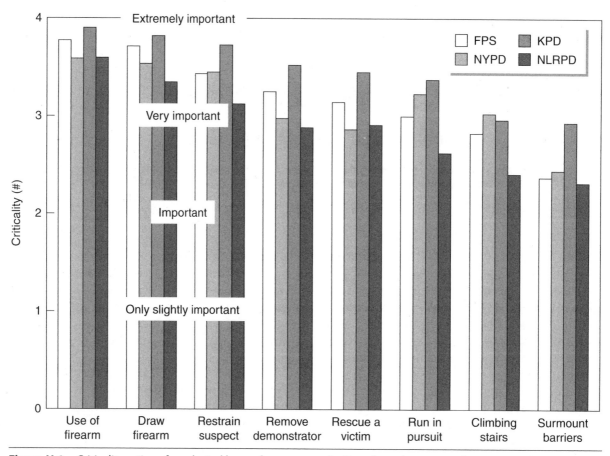

Figure H.1 Criticality ratings for selected law enforcement tasks. Based on a large, a medium-size, and a small department and a federal law enforcement agency.

Data from L. Cunningham, 2006, Physical performance constructs: Attributes of an effective police officer. In *Hard work: Physically demanding occupations* [a symposium presented at the annual meeting of the American College of Sports Medicine, Denver].

for the participation of fast-twitch, anaerobic types of muscle fibers. Performance on a law enforcement physical abilities test is highly related to aerobic power (Cunningham 2006). But regardless of the physical ability of the officer, it is the physical characteristics of the offender that define the mission.

The second major area affecting physical abilities under the banner of essential functions is struggling with someone who does not wish to be constrained or, even worse, wishes to inflict physical harm on the officer. It is one thing to match the velocity of a fleeing felon, but, like the dog that chases the automobile, what do you do when you catch him? There is limited scientific literature identifying predictors of success in physical struggles. What we know from the world of sport is that physical size (e.g., lean body weight) dominates factors that emerge in a multivariant analysis. In the world of sport, competition is determined by weight classes; rarely will boxers or wrestlers compete above their weight class.

Tests of muscular fitness, including strength (defined as maximum force), muscular endurance (repetitions of a submaximal resistance), and power (explosive force or a sprint), are associated with success in law enforcement tasks. Data derived from a battery of physical ability tests are significantly related to performance on criterion tasks, tasks that are routinely documented by police officers across a wide range of jurisdictions.

In case there is any remaining doubt about the occurrence of such tasks as scaling walls, entering windows, dragging or lifting bodies, crawling under or over an assortment of obstacles, and restraining and obtaining compliance in resisting perpetrators, data collected by Davis, Cunningham, and many others consistently reveal a high level of agreement on the performance of these tasks. These data have been derived from rural counties, small and large cities, and federal organizations such as the FBI.

After hundreds of JTAs in a variety of organizations, there is a clear trend toward the identification of a core set of essential functions that are common to law enforcement officers throughout the country. Journeyman law enforcement officers practice their craft employing the same set of skills and the same muscular and energy pathways. A similar set of essential functions exists for structural firefighters (see figure H.2).

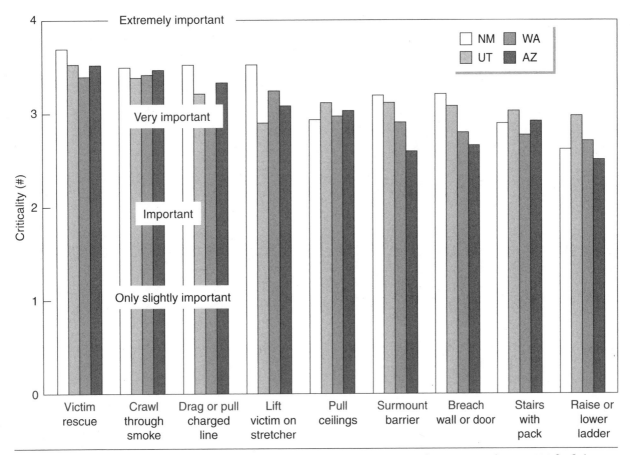

Figure H.2 Criticality ratings for selected firefighting tasks. For departments in four states with 60 to 300 firefighters per department.

Data from L. Cunningham, 2006, Physical performance constructs: Attributes of an effective police officer. In *Hard work: Physically demanding occupations* [a symposium presented at the annual meeting of the American College of Sports Medicine, Denver].

Appendix I

Assessment of Performance During Manual Timber Harvesting

Delia Roberts, PhD • Selkirk College, British Columbia

Employees perform physical work under rigorous environmental conditions in many occupations. In many cases, because access to the work site is difficult, it is desirable to extend the work shift for as long as possible in the interests of efficiency. However, an important consideration limiting shift length is that fatigue contributes significantly to decreased worker productivity and the incidence of injury. This becomes especially important in occupations with a high skill level and in which a loss of mental focus or coordination can lead to serious bodily harm, such as in first responders, wildland firefighting, and manual timber harvesting.

In spite of difficult logistical issues, it is important to accurately and reliably measure work output in workers in the field. Only in this manner can we begin to answer the question of when fatigue actually compromises worker productivity and safety. Following are some of the questions that must be addressed:

- What are the demands of the task?
- How does worker productivity change as shift length is extended and workers become fatigued?
- How are injury and accident rates related to the extension of work shifts and the development of fatigue?

This investigation began by determining the work output and physiological demands of workers engaged in manual falling and bucking of trees in remote and mountainous terrain. Access was by helicopter only.

Data were collected on 10 male fallers and buckers (mean±SD age, height, and weight were 37.9±7.6 yrs, 178.6±7.4 cm, and 86.0±8.7 kg, respectively) on two consecutive days, once a month for a three-month period. Heart rates were recorded every 15 seconds by telemetry (Polar S-610), and activity levels were recorded using triaxial accelerometers (Actical system, Mini-Mitter), mounted at the right anterior superior iliac crest. In addition, sweat rates (forearm patches), dietary fluid, and urinary output were also monitored.

Loggers worked a mean of 6.3±1.9 hours at a mean heart rate of 100±10.8 beats per minute with 3.7±3.0 hours spent at intensities greater than 40 percent $\dot{V}O_2$max. The mean activity count during work was 518±79 cpm. Body mass showed significant

Table I.1 Temperature, Humidity, and Sweat Loss

Mean ±	June	July	August
Ambient temperature (°C)	9.6±2.2	19.3±1.9	16.9±2.5
Relative humidity (%)	75.1±10.5	80.2±9.9	74.9±2.2
Daily mass loss (%)	2.1±1.0	2.1±0.9	1.7±0.7
Daily fluid intake (L)	1.9±0.7	2.1±0.7	1.8±0.7
Estimated daily sweat loss (L)	3.0±1.4	3.3±1.2	2.5±0.9

losses on a daily basis pre- to postwork with recovery by the following morning, indicating the significant generation of metabolic heat in spite of cool ambient temperatures.

No direct measure of productivity was available. The most obvious outcome, volume of wood cut per unit of time, did not reflect the degree of movement difficulty associated with what was, in some cases, very steep and hazardous terrain. Nor were researchers able to discern how much bucking was required to clear the main tree trunks of branches. The time to harvest a 5 meter × 10 meter plot was similarly an unsuitable measurement. In contrast, the number of seconds in a three-minute period spent with the saw blade engaged (discernable by the change in tone as the blade bit into the wood) appeared to be sensitive to fatigue (see figure I.1).

The apparent decline in productivity immediately prior to lunch and at the end of the day may have been due to fatigue. That cutting increased again following an hour-long break with substantial food intake suggests that this was indeed the case. Further work is required to determine the cause of the fatigue and the implications for health and safety. Data on wildland firefighters indicate that performance is maintained when liquid and or solid carbohydrate supplements are consumed throughout the workday (Ruby et al. 2004).

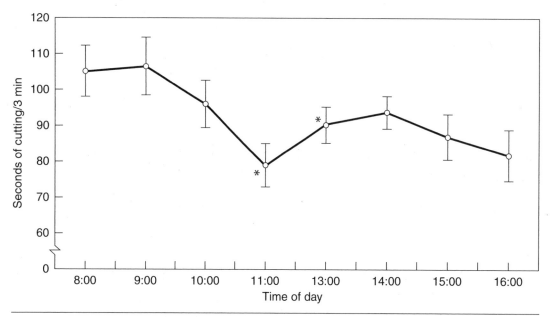

Figure I.1   Productivity as measured by the number of seconds of cutting in a three-minute period.

*Significantly different from the previous point, $p < 0.05$.

References

Age Discrimination in Employment Act of 1967, 29 U.S.C. §§ 62 et seq. (2006).

Alexandria Volunteer Fire Dept. v. Rule, 725 F.2d 673 (4th Cir. 1984).

American College of Sports Medicine. 1998. Exercise and physical activity for older adults. *Medicine & Science in Sports & Exercise* 30: 992-1008.

American College of Sports Medicine. 2002. *Physical training for improved occupational performance.* Indianapolis, IN: Current Comment. www.acsm.org.

American College of Sports Medicine. 2005. *ACSM's guidelines for exercise testing and prescription. 7th ed.* Philadelphia: Lippincott Williams & Wilkins.

American College of Sports Medicine. 2007. Exercise and acute cardiovascular events: Placing the risks into perspective. *Medicine & Science in Sports & Exercise,* 39:886-901.

American Heart Association. 2007. Physical activity. www.americanheart.org/presenter.jhtml?identifier=4563.

American Industrial Hygiene Association. 2006. Z88.6 Respirator: Physical Qualifications for Personnel. Fairfax, VA:AIHA. www.aiha.org/content/insideaiha/standards/z88.htm/.

Americans with Disabilities Act of 1990, 42 U.S.C. §§ 12101 et seq. (2006).

Amoroso, P., Reynolds, K., Barnes, D., and White, D. 1996. *Tobacco and injury.* Natick, MA: U.S. Army Research Institute for Environmental Medicine, Technical Report 96-1.

Armstrong, L. 2000. *Performing in extreme environments.* Champaign, IL: Human Kinetics.

Armstrong, L., Casa, D., Millard-Stafford, M., Moran, D., Pyne, S., and Roberts, W. 2007. Exertional heat illness during training and competition. *Medicine & Science in Sports & Exercise* 39: 556-572.

Astrand, P.O., Rodahl, K., Dahl, H., and Stromme, S. 2003. *Textbook of work physiology.* Champaign, IL: Human Kinetics.

Bachop, Steve. 2001. Personal communication.

Balke, B. 1963. *A simple field test for the assessment of physical fitness.* Report no. 63-6. Oklahoma City: Civil Aeronautic Research Institute, Federal Aviation Agency.

Baris, D., Garrity, T., Telles, J., Heineman, E., and Hoar Zahm, S. 2000. A cohort mortality study of Philadelphia firefighters. *Fire Engineering,* February.

Barnard, R., Gardner, G., Diaco, N., and Kattus, A. 1972. *Ischemic response to sudden strenuous exercise.* Paper presented at the annual meeting of the American College of Sports Medicine, Philadelphia.

Beckett, M.B., and Hodgdon, J.A. 1987. *Lifting and carrying capacities relative to physical fitness measures* (Report No. 87-26). San Diego, CA: Naval Health Research Center.

Bell, G., Syrotuik, D., Martin, T., Burham, R., and Quinney, H. 2000. Effects of concurrent strength and endurance training on skeletal muscle properties and hormone concentrations in humans. *Journal of Applied Physiology* 81: 418-427.

Bernaards, C., Jans, M., vandenHeuvel, S., Hendriksen, I., Houtman, I., and Bongers, P. 2006. Can strenuous leisure time physical activity prevent psychological complaints in a working population? *Occupational and Environmental Medicine* 63: 10-16.

Blackburn, H. 1980. Risk factors and cardiovascular disease. *The American Heart Association Heart Book.* New York: E.P. Dutton.

Bouchard, C., An, P., Rice, T., Skinner, J., Wilmore, J., Gagnon, J., Perusse, L., Leon, A., and Rao, D. 1999. Familial aggregation of V̇O₂max response to exercise training: Results from the Heritage Family Study. *Journal of Applied Physiology* 87: 1003-1008.

Bouchard, C., Boulay, M., Simoneau, J., Lorrie, G., and Pierrise, L. 1988. Heritability of aerobic and anaerobic performance: An update. *Sports Medicine* 5: 69-73.

British Columbia (Public Service Employee Relations Commission) v. BC Government Service Employees Union, File #26274, 1999.

Brogden, H.E. 1949. When testing pays off. *Personnel Psychology* 2: 171-183.

Bronstein, D. 1993. *Law for the expert witness.* Boston: Lewis.

Budd, G., Brotherhood, J., Hendrie, L. Cheney, P., and Dawson, M. 1996. Safe and productive bushfire fighting with hand tools. Canberra: Australian Government Publishing Service.

Budd, G., Brotherhood, J., Hendrie, A., Jeffery, S., Beasley, F., Costin, B., Zhien, W., Baker, M., Cheney, N., and Dawson, N. 1997. Physiological and subjective responses of men suppressing wildland fires. *International Journal of Wildland Fire* 7:133-145.

Buddin, R.J. 2005. Success of first-term soldiers: The effects of recruiting practices and recruit characteristics. RAND Corporation. www.rand.org/pubs/monographs/2005/RAND_MG262.pdf.

Cady, L., Bishoff, D., O'Connell, E., Thomas, P., and Allan, J. 1979. Back injuries in firefighters. *Journal of Occupational Medicine* 21: 269-272.

Cady, L., Bischoff, D., O'Connell, E., Thomas, P., and Allen, J. 1979. Back injuries in firefighters. *Journal of Occupational Medicine* 21: 269-272.

Canadian Centre for Occupational Health and Safety. 2002. OSH answers: General practice. www.ccohs.ca/oshanswers/ergonomics/mmh/generalpractice.html.

Carpenter, C., and Nelson, B. 1999. Low back strengthening for the prevention of low back pain. *Medicine & Science in Sports & Exercise* 31: 18-24.

Carpenter, R.A., Finley, C., and The Cooper Institute. 2005. *Healthy Eating Every Day.* Champaign, IL: Human Kinetics.

Castellani, J., Young, A., Ducharme, M., Giesbrecht, G., Glickman, E., and Sallis, R. 2006. Prevention of cold injuries during exercise. *Medicine & Science in Sports & Exercise* 38: 2012-2029.

Cattin, P. 1978. The predictive validity based procedure for choosing between regression and equal weights. *Organizational behavior and human performance* 22: 93-102.

Centers for Disease Control and Prevention. 1999. Framework for program evaluation in public health. *MMWR*, 48:RR-11.

Chapman, R., Stray-Gunderson, J., and Levine, B. 1998. Individual variation in response to altitude training. *Journal of Applied Physiology* 85: 1448-1456.

Chenoweth, D. 1998. *Worksite health promotion*. Champaign, IL: Human Kinetics.

Civil Rights Act of 1964, Title VII, 42 U.S.C. 77 2000e et seq. (2006).

Civil Service Reform Act of 1978, U.S.C. 2302(B)(5) (2006).

Cole, D., Ibrahim, S., Shannon, H., Scott, F., and Eyles, J. 2001. Work correlates of back problems and activity restriction due to musculoskeletal disorders in the Canadian national population health survey (NPHS) 1994-5 data. *Occupational and Environmental Medicine*, 58:728-734.

Cooper Institute. 2007. Common questions regarding physical fitness tests, standards and programs for public safety. www.cooperinst.org/education/law_enforcement/documents/Fitness_Questions.pdf.

Cooper, K. 1968. *Aerobics*. New York: Bantam Books.

Corbin, C. 2005. *Fitness for Life Wellness DVD*. Champaign, IL: Human Kinetics.

Cordes, K., and Sharkey, B. 1995. Physiological comparison of protective clothing variations. *Medicine & Science in Sports & Exercise* 27: s279.

Cox, C. 2003. *ACSM's worksite health promotion manual*. Champaign, IL: Human Kinetics.

Coyle, E., Hemmert, M., and Coggan, A. 1986. Effects of detraining on cardiovascular responses to exercise: Role of blood volume. *Journal of Applied Physiology* 60: 95-99.

Cunningham, L. 2006. Physical performance constructs: Attributes of an effective police officer. In *Hard work: Physically demanding occupations*, a symposium presented at the annual meeting of the American College of Sports Medicine, Denver.

Davis, Paul O. Testimony in *Lanning v. Southeast Pennsylvania Transit Authority* (SEPTA). Civil action 97-0593, civil action 97-1161.

Davis, P., Crumling, H., Sharkey, B., Zung V., Tran, Z., and O'Connor, J. 1994. *Development of a physical performance test for the Federal Bureau of Investigation*. Burtonsville, MD: Applied Research Associates.

Davis, P., Dotson, C., and Santa Maria, D. Laine. 1982. Relationship between simulated firefighting tasks and physical performance measures. *Medicine and Science in Sports and Exercise* 14 (1): 65-71.

Davis, P., Dotson, C., and Sharkey, B. 1986. *Physical fitness requirements of United States Marine Corps*. Langley Park, MD: Institute of Human Performance.

Davis, P., and Starck, A. 1980. Age and performance in a police population. *Law Enforcement Bulletin*. Washington, DC: Federal Bureau of Investigation, 15-21.

Dawes, R.M., and Corrigan, B. 1974. Linear models in decision-making. *Psychological Bulletin* 81: 95-106.

Debeliso, M., O'Shea, J., Harris, C., Adams, K., and Climstein, M. 2004. The relationship between trunk strength measures and lumbar disc deformation during stoop type lifting. *Journal of Exercise Physiology* (online) 7: 16-26.

DeLorenzo-Green, T., and Sharkey, B. 1995. Development and validation of a work capacity test for wildland firefighters. *Medicine & Science in Sports & Exercise* 27: s166.

Des Moines v. Civil Service Commission, 540 N.W.2d 52 (Iowa 1995).

Di Vico, P. 2006. Physical performance constructs: Attributes of an effective structural firefighter. In *Hard work: Physically demanding occupations*, a symposium presented at the annual meeting of the American College of Sports Medicine, Denver.

Di Vico, P., Cunningham, L., Armeli, T., and Davis, P. 2004. *Job task analysis, job standard test validation and physical fitness standards development: New York City Police Department*. Richland, WA: Health Metrics.

Di Vico, P., Cunningham, L., Armeli, T., and Davis, P. 2005. Construct valid approach to development of a physical ability test for law enforcement personnel. *Medicine and Science in Sports and Exercise* 37 (5): s403-4.

Docherty, D., McFadyen, P., and Sleivert, G. 1992. *Bona fide occupational fitness tests and standards for B.C. Forest Service wildland firefighters*. Victoria, BC: University of Victoria.

Dothard v. Rawlinson (433 U.S. 321, 97 S.Ct. 2720), 1977.

Dotson, C., Santa Maria, D., and Davis, P. 1976. Development of a job-related physical performance examination for firefighters. Washington DC: U.S. Department of Commerce, National Fire Prevention and Control Administration.

Duncan, G., Li, S., and Zhou, X. 2005. Cardiovascular fitness among U.S. adults: NHANES 1999-2000 and 2001-2002 . *Medicine & Science in Sports & Exercise* 37: 1324-1328.

Eberle v. State of Missouri, 779 S.W.2d 302 (Mo.App. W.D.1989).

EEOC v. Simpson Timber Co. DC, W.Wash. C89-1455 WD 1992.

EEOC. October 1995. ADA enforcement guidance: Preemployment disability-related questions and medical examinations. www.eeoc.gov/policy/docs/preemp.html.

Engholm, G., and Holmstrom, E. 2006. Dose-response associations between musculoskeletal disorders and physical and psychosocial factors among construction workers. *Scandinavian Journal of Work Environment and Health* 31: 57-67.

Equal Employment Opportunity Commission. (1992). *Technical assistance manual for the Americans with Disabilities Act*. Washington, DC: Warren Gorham Lamont.

Equal Pay Act of 1963, 29 U.S.C. § 206 (2006).

Eriksen, W., Natvig, B., and Bruusgaard, D. 1999. Smoking, heavy physical work and low back pain: A four-year prospective study. *Occupational Medicine* 49: 155-160.

Fahy, R. 2005. *U.S. firefighter fatalities due to sudden cardiac death, 1995-2005*. Quincy, MA: National Fire Protection Association. www.nfpa.org.

Federal Register. (1978). *Uniform guidelines on employee selection procedures*, 43, 38290-38315.

Ferrante v. Niagara County, 551 N.Y.S. 2d 134 (N.Y.A.D.4Dept. 1990).

Feuerstein, M., Berkowitz, S., Haufler, A., Lopez, M., and Huang, G. 2001. Working with low back pain: Workplace and individual psychosocial determinants of limited duty and lost time. *American Journal of Industrial Medicine* 40: 626-638.

Fiatarone, M., O'Neill, E., Doyle-Ryan, N., Clements, K., Solares, G., Nelson, M., Roberts, S., Kehayias, J., Lipsitz, L., and Evans, W. 1994. Exercise training and nutritional supplementation for physical frailty in very elderly people. *New England Journal of Medicine* 330: 1769-1775.

Fleishman, E., and Mumford, M. 1988. Ability requirement scales. In *The job analysis handbook for business, industry, and government*, ed. S. Gael, 917-935. New York: John Wiley & Sons.

Fletcher, G., Balady, G., Froelicher, V., Hartley, H., Haskell, W., and Pollock, M. 1995. Exercise standards: A statement for healthcare professionals from the American Heart Association. *Circulation* 91: 580-615.

Foster, J., and Porcari, J. 2001. The risks of exercise training. *Journal of Cardiopulmonary Rehabilitation* 21: 347-352.

Franco, O., deLaet, C., Peeters, A., Jonker, J., Mackenbach, J., and Nusselder, W. 2005. Effects of physical activity on life expectancy with cardiovascular disease. *Archives of Internal Medicine* 165: 2355-2360.

Gaffin, S., and Hubbard, R. 2001. Pathophysiology of heat stroke. In *Medical aspects of harsh environments*, eds. K. Pandolf and R. Burr. Washington, DC: U.S. Government Printing Office.

Gaines, L.K., Fulkenberg, S., and Gambino, J. (1993). Police physical agility testing: A historical and legal analysis, *American Journal of Police* 12 (4): 47-67.

Ganzel, D. 2002. Physical standards for police cyclists. *IPMBA News*. www.ipmba.org/newsletter-0210-phys.htm.

Gardner, J. 2002. Death by water intoxification. *Military Medicine* 167: 306-311.

Gaskill, S., and Ruby, B. 2004. Relationship of work to salivary IgA and fatigue in wildland firefighters. *Wildland Firefighter Health and Safety Report* 10: 8.

Gaskill, S., Ruby, B., Heil, D., Sharkey, B., Hansen, K., and Lankford, D. 2002. Fitness, work rates and fatigue during arduous wildfire suppression. *Medicine & Science in Sports & Exercise* 34: s195.

Gaul, K. 2000. *Issues of gender and size in emergency response occupations*. Paper presented at the annual meeting of the Canadian Society of Exercise Physiology, Canmore, Alberta.

Gibbons, R., Balady, G., Bricker, J., Chaaitman, B. Fletcher, G., Froelicher, V., Mark, D., McCallister, B., Mooss, A., O'Reilly, M., and Winters, W. 2002. Guideline update for exercise testing: A report of the American College of Cardiology/American Heart Association Task Force on Practice Guidelines. Bethesda, MD: American College of Cardiology. www.acc.org/clinical/guidelines/exercise/dirIndex.htm/.

Gledhill, N., Warburton, D., and Jamnik, V. 1999. Haemoglobin, blood volume, cardiac function, and aerobic power. *Canadian Journal of Applied Physiology* 24: 54-65.

Gomez-Merino, D., Chennoui, M., Burnat, P., Drogou, C., and Guezennec, C. 2003. Immune and hormonal changes following intense military training. *Military Medicine* 168: 1034-1038.

Guion, R.M. 1998. *Assessment, measurement, and prediction for personnel decisions*. Mahwah, NJ: Erlbaum Associates.

Gutekunst, D., Harmen, E., and Frykman, P. 2005. Determinants of 2 mile (3.2 km) time with and without a 70 pound (32 kg) military load. *Medicine & Science in Sports & Exercise* 37: s403.

Harman, E., Frykman, P., Lammi, E., and Palmer, C. 1997. *Effects of a specifically designed physical conditioning program on the load carriage and lifting performance of female soldiers*. Natick, MA: U.S. Army Research Institute of Environmental Medicine.

Hartman, B., and Fleischer, A. 2005. Physical load exposure at construction sites. *Scandinavian Journal of Work Environment and Health* 31: 88-95.

Haskell, W., Lee, I., Pate, R., Powell, K., Blair, S., Franklin, B., Macera, C., Heath, G., Thompson, P., and Bauman, A. 2007. Physical activity and public health: Updated recommendation for adults from the American College of Sports Medicine and the American Heart Association. *Medicine & Science in Sports & Exercise*, 39:1423-1434.

Henderson, N., Berry, M., and Matic, T. 2007. Field measures of strength and fitness predict firefighter performance on physically demanding tasks. *Personnel Psychology*, 60:431-473.

Hickson, R. 1980. Interference of strength development by simultaneously training for strength and endurance. *Journal of Applied Physiology* 45: 255-263.

Hodgdon, J. 1992. Body composition in the military services: Standards and methods. In *Body composition and physical performance: Applications to the military services*, eds. B. Marriott and J. Grumstrup-Scott, 57-70. Washington, DC: National Academy Press.

Hogan, J. 1991. Structure of physical performance in occupational tasks. *Journal of Applied Psychology* 76: 495-507.

Hogan, J., and Quigley, A. 1986. Physical standards for employment and courts. *American Psychologist* 41: 1193-1217.

Hollar, D. 2000. Physical ability tests and Title VII. *University of Chicago Law Review. 67U.Chi.L.Rev. 777*.

Holloszy, J., Dalsky, G., Nemeth, P., Hurley, B., Martin, W., and Hagberg, J. 1986. Utilization of fat as a substrate during exercise: Effect of training. In *Biochemistry of exercise IV*, ed. B. Saltin, 183-190. Champaign, IL: Human Kinetics.

Hood, D., Takahashi, M., Connor, M., and Freyssenet, D. 2000. Assembly of the cellular powerhouse: Current issues in muscle mitochondrial biogenesis. *Exercise and Sports Science Reviews* 28: 68-73.

Hubal, M., Gordish-Dressman, H., Thompson, P., Price, T., Hoffman, E., Angelopolous, T., Gordon, P., Moyna, N., Pescatello, L., Visich, P., Zoeller, R., Seip, R., and Clarkson, P. 2005. Variability in muscle size and strength gain after unilateral resistance training. *Medicine & Science in Sports & Exercise* 37: 961-972.

Human Rights Commission State of Minnesota vs St. Paul Fire Department. 1989. OAH Docket No. 8-1700-3224-2.

Hunter, J.E., Schmidt, F.L., and Judiesch, M.K. 1990. Individual differences in output variability as a function of job complexity. *Journal of Applied Psychology* 75: 28-42.

Institute of Medicine. 2002. Dietary Reference Intake, Energy, Carbohydrate, Fiber, Fat, Fatty Acids, Cholesterol, Protein, and Amino Acids. Washington D.C., National Academy Press. www.nap.edu/books/0309085373/html/.

International Association of Fire Fighters. (2007). The IAFF/ICH-IEFS Fire Service joint labor management candidate physical ability test program summary. www.iaff.org/HS/CPAT/cpat_index.html.

Jackson, A. 1994. Preemployment physical evaluation. *Exercise and Sport Science Review* 22: 53-90.

Jensen, K. 2005. *Federal interagency wildland firefighter medical qualification standards*. Boise, ID: National Wildfire Coordinating Group.

Kales, S., Soteriades, E., Christophi, C., and Christiani, D. 2007. Emergency duties and deaths from heart disease among firefighters in the United States. *New England Journal of Medicine* 356: 1207-1215.

Kasch, F., Boyer, J., Van Camp, S., Verity, L., and Wallace, J. 1990. The effects of physical activity and inactivity on aerobic power in older men: A longitudinal study. *Physician and Sportsmedicine*, 18:73-83.

Kasch, F. 2001. Thirty-three years of aerobic exercise adherence. *Quest* 53: 362-365.

Kasch, F., Wallace, J., Van Camp, S., and Verity, L. 1988. A longitudinal study of cardiovascular stability in active men aged 45-65 years. *Physician and Sportsmedicine* 16: 117-126.

Klamath Falls Fire Department v. KF Firefighters (Arbitration) 1996.

Klissouras, V. 1976. Heritability of adaptive variation. *Journal of Applied Physiology* 31: 338-344.

Knapik, J. 1997. The influence of physical fitness training on the manual materials handling capability of women. *Applied Ergonomics* 28: 339-345.

Knapik, J., Darakjy, S., Hauret, K., Canada, S., Scott, S., Rieger, W., Marin, R., and Jones, B. 2006. Increasing the physical fitness of low-fit recruits before basic combat training: An evaluation of fitness, injuries, and training outcomes. *Military Medicine* 171: 45-54.

Knapik, J., Hauret, H., Arnold, S., Canham-Chervak, M., Mansfield, A., Hoedebecke, E., and McMillian, D. 2003. Injury and fitness outcomes during implementation of physical readiness training. *International Journal of Sports Medicine* 24: 372-381.

Kramer, W., Mazzetti, S. Nindl, B., Gotshalk, L., Volek, J., Bush, J., Marx, J., Dohi, K., Gomez, A., Miles, M., Fleck, S., Newton, R., and Hakkinen, K. 2001. Effect of resistance training on women's strength/power and occupational performances. *Medicine & Science in Sports & Exercise* 33: 1011-1025.

Landy, F. 1992. *Alternatives to chronological age in determining standards of suitability for public safety jobs.* Volume 1, Technical Report, State College, PA: Pennsylvania State University.

Lanning v. SEPTA, 181 F.3d 478 (3d Cir. 1998).

Lanning v. SEPTA, 2000 U.S. Dist. LEXIS 17612 (Dec. 7, 2000).

Lanning v. SEPTA, 308 F.3d 286 (3d Cir. 2002).

Legault v. aRusso, 842 F.Supp. 1479 (D.N.H. 1994).

Lindquist, C., and Bray, R. 2001. Trends in overweight and physical activity among U.S. military personnel, 1995-1998. *Preventive Medicine* 32: 57-65.

Lord, F.M. 1962. Cutting scores and errors of measurement. *Psychometrica* 27: 19-30.

Lord, F.M. 1963. Cutting scores and errors of measurement: A second case. *Educational and Psychological Measurement* 23: 63-68.

Malchaire, J., Roquelaure, Y., Cock, N., Piette, A., Vergracht, S., and Chiron, H. 2002. Musculoskeletal complaints, functional capacity, personality and psychosocial factors. *International Archives of Occupational and Environmental Health* 74: 549-557.

Maron, B., Thompson, P., Puffer, J., McGrew, C., Strong, W., Douglas, P., Clark, L., Mitten, M., Crawford, M., Atkins, D., Driscoll, D., and Epstein, E. 1996. Cardiovascular preparticipation screening of competitive athletes. *Circulation* 94: 850-856.

McCormick, E.J. 1979. *Job analysis: Methods and applications.* New York: Amacom.

McSweeney, K., Congleton, J., Kerk, C. Jenkins, O., and Craig, B. 1999. Correlation of recorded injury and illness data with smoking, exercise, and absolute aerobic capacity. *International Journal of Industrial Ergonomics* 24: 193-200.

Meridian v. Firefighters Association of Michigan, MERC #C95-H174, 9 MPER 1996.

Miller, A., MacDougall, J., Tarnopolsky, M., and Sale, D. 1993. Gender differences in strength and muscle fiber characteristics. *European Journal of Applied Physiology and Occupational Physiology* 66: 254-262.

Moran, D., Shitzer, A., and Pandolf, K. 1998. A physiological strain index to evaluate heat stress. *American Journal of Physiology* 275: R129-R134.

National Center for Health Statistics (NCHS). 2003-04. Prevalence of Overweight and Obesity Among Adults: United States, 2003-2004. National Health and Nutrition Examination Survey Data. www.cdc.gov/nchs/products/pubs/pubd/hestats/overweight/overwght_adult_03.htm.

National Fallen Firefighters Foundation. 2005. Report of the National Fire Service research agenda symposium. www.everyonegoeshome.com/report.pdf.

National Institute for Occupational Safety and Health (NIOSH). June 2007. Preventing fire fighter fatalities due to heart attacks and other sudden cardiovascular events. Publication No. 2007-133. www.cdc.gov/niosh/docs/2007-133.

Nevola, V., Collins, S., Puxley, K., Bentley, M., and Withey, W. 2004. Aerobic demands of conducting a simulated footchase, arrest and handcuff task in the UK police. *Medicine & Science in Sports & Exercise* 36: s245.

NFPA 1583. 2000. Standard on Health-Related Fitness Programs for Fire Fighters, Quincy, MA: National Fire Protection Association.

NFPA 1582. 2000. Standard on Medical Requirements for Fire Fighters and Information for Fire Department Physicians, Quincy, MA: National Fire Protection Association.

NFPA 1500. 2002. Standard on Fire Department Occupational Safety and Health Program, Quincey, MA: National Fire Protection Association.

OSHA 2002. Respiratory Protection Program. Washington, D.C.: Occupational Safety and Health Administration, (29 CFR 1910.134) (2002).

Panel on Energy, Obesity and Body Weight Standards, 1987, *American Journal of Clinical Nutrition* 45: 1035-1041.

Passmore, R., and Durnin, J. 1955. Human energy expenditure. *Physiological Review* 35: 801-824.

Pate, R., Pratt, M., Blair, S., Haskell, W., Macera, C., Bouchard, C., Buchner, D., Ettinger, W., Heath, G., King, A., Kriska, A., Leon, A., Marcus, B., Morris, J., Paffenbarger, R., Patrick, K., Pollock, M., Rippe, J., Sallis, J., and Wilmore, J. 1995. Physical activity and public health: A recommendation from the Centers for Disease Control and Prevention and the American College of Sports Medicine. *Journal of the American Medical Association* 273: 402-407.

Patrolmen's Benevolent Association v. Township of East Brunswick, 433 A.2d 183 N.J.Super A.D. 1984).

Personnel Administrator of Massachusetts v. Feeney, 442 U.S. 256 (1979).

Pentagon Force Protection Agency v. Fraternal Order of Police DPS, Labor Committee (2004).

Peate, W., Lundergan, L., and Johnson, J. 2002. Fitness self-perception and $\dot{V}O_2$max in firefighters. *Journal of Occupational and Environmental Medicine* 44: 546-550.

Pierce v. Franklin Electric Co., 737 P.2d 921 (Okla. 1987).

Pimentel, A., Gentile, H., and Tanaka, D. 2003. Greater rate of decline in maximal aerobic capacity with age in endurance-trained than in sedentary men. *Journal of Applied Physiology* 94: 2406-2413.

Powel, K., and R. Paffenbarger. 1985. Workshop on epidemiologic and public health aspects of physical activity and exercise: A summary. *Public Health Reports* 100: 118-126.

Pronk, N. 2003. A sense of urgency to improve employee health: The bottom line and the role of worksite health promotion. *ACSM's Health and Fitness Journal* 7: 6-11.

Pyne, S. 2001. *Fire: A Brief History.* Seattle: University of Washington Press.

Rehabilitation Act of 1973, 29 U.S.C. §§ 701 et seq. (2006).

Reilly, T., and Freeman, K. 2006. Effect of loading on spinal shrinkage in males of different age groups. *Applied Ergonomics* 37: 305-310.

Reinhardt, T., and Ottmar, R. 1997. Employee exposure review. In *Health hazards of smoke: Recommendations of the consensus conference,* ed. B. Sharkey. Missoula, MT: USDA Forest Service, Technology and Development.

Roberts, D. 2002. Physiological and biochemical status of reforestation workers. *Journal of Occupational and Environmental Medicine* 44: 559-567.

Roberts, D. 2004. Exceptionally high plasma cortisol and IL-6 levels in reforestation workers. *Medicine & Science in Sports & Exercise* 36: s220.

Roberts, D. 2006. Training for tree planting. Personal communication.

Roberts, D. 2006. Personal communication.

Roth, E. 1968. *Compendium of human responses to the aerospace environment III.* Washington DC, National Aeronautics and Space Administration.

Rothwell, T., and Sharkey, B. 1996. The effect of an air-purifying respirator on performance of upper body work. *Medicine & Science in Sports & Exercise* 28: s87.

Round, A., and Green, D. 1998. Can money motivate firefighters to exercise? *Fire Engineering,* September: 119-122.

Ruby, B., Gaskill, S., Harger, D., Heil, D., and Sharkey, B. 2004. *Carbohydrate feedings increase self-selected work rates during arduous wildfire suppression.* Paper presented at the annual meeting of the American College of Sports Medicine, Indianapolis, IN.

Ruby, B., Ledbetter, G., Armstrong, D., and Gaskill, S. 2003. Wildland firefighter load carriage: Effects on transit time and physiological responses during escape to safety zone. *International Journal of Wildland Fire* 12: 111-116.

Ruby, B., Shriver, T., Zedric, T., Sharkey, B., Burks, C., and Tysk, S. 2002. Total energy expenditure during arduous wildfire suppression. *Medicine & Science in Sports & Exercise* 34: 1048-1054.

Sackett, P., and Mavor, A., eds. 2006. *Assessing fitness for military enlistment.* Washington, DC: National Research Council.

Sarno, M. 2003. An analysis of *Lanning v. Southeastern Pennsylvania Transportation Authority* and the business necessity defense as applied in third circuit discrimination cases. *Villanova Law Review.48 Vill.L.Rev. 1403.*

Schmidt, F.L., and Hunter, J.E. 1983. Individual differences in productivity: An empirical test of estimates derived from studies of selection procedure utility. *Journal of Applied Psychology* 68: 407-415.

Schmidt, F.L., and Hunter, J.E. 1998. The validity and utility of selection methods in personnel psychology: Practical and theoretical implications of 85 years of research findings. *Psychological Bulletin* 124: 262-274.

Schmidt, F.L., Hunter, J.E., McKenzie, R.C., and Muldrow, T.W. 1979. The impact of valid selection procedures on work-force productivity. *Journal of Applied Psychology* 64: 609-626.

Schmidt, F.L., Hunter, J.E., and Pearlman, K. 1982. Assessing the economic impact of personnel programs on workforce productivity. *Personnel Psychology* 35: 333-347.

Selye, H. (1985). The nature of stress. *Basal Facts* 7 (1): 3-11.

Sharkey, B. 1977. *Fitness and work capacity.* Washington, DC: U.S. Government Printing Office.

Sharkey, B. 1981. Fitness for wildland firefighting. *Physician and Sportsmedicine* 9: 77-83.

Sharkey, B. 1984. *Training for cross-country ski racing.* Champaign, IL: Human Kinetics.

Sharkey, B. 1987. Functional vs. chronological age. *Medicine & Science in Sports & Exercise* 19: 174-178.

Sharkey, B. 1990. *Physiology of fitness.* Champaign, IL: Human Kinetics.

Sharkey, B. 1991. *New dimensions in aerobic fitness.* Champaign, IL: Human Kinetics.

Sharkey, B. 1997a. *The development and validation of a job-related work capacity test for wildland firefighting.* Paper presented at the annual meeting of the American College of Sports Medicine, Denver, CO.

Sharkey, B. 1997b. *Fitness and work capacity.* 2nd ed. Missoula, MT: USDA Forest Service.

Sharkey, B. 1997c. Respiratory protection. In *Health hazards of smoke: Recommendations of the consensus conference,* ed. B. Sharkey. Missoula, MT: USDA Forest Service, Technology and Development.

Sharkey, B., and Gaskill, S. 2006. *Sport physiology for coaches.* Champaign, IL: Human Kinetics.

Sharkey, B., and Gaskill, S. 2007. *Fitness and health.* 6th ed. Champaign, IL: Human Kinetics.

Sharkey, B., Jukkala, A., and Herzberg, R. 1984. *Fitness trail.* Missoula, MT: USDA Forest Service Technology and Development Center.

Sharkey, B., and Mead, Z. 1993. Predicting the effect of an air-purifying respirator on work performance. *Medicine & Science in Sports & Exercise* 25: s120.

Sharkey, B., Rothwell, T., and DeLorenzo-Green, T. 1994. Development of a job-related work capacity test for wildland firefighters. *Medicine & Science in Sports & Exercise* 26: s88.

Sharkey, B., Rothwell, T., and Jukkala, A. 1996. Validation and field evaluation of a work capacity test for wildland firefighters. *Medicine & Science in Sports & Exercise* 28: s79.

Sharp, M. 1994. Physical fitness and occupational performance of women in the U.S. Army. *Work* 4: 80-92.

Shell, D. 2003. Exercise science and law enforcement. *Strength and Conditioning Journal* 25 (3), 52-57.

Shephard, R., and Bonneau, J. 2002. Assuring gender equity in recruitment standards for police officers. *Canadian Journal of Applied Physiology* 27: 263-295.

Siscovick, D., LaPorte, R., and Newman, J. 1985. The disease-specific benefits and risks of physical activity and exercise. *Public Health Reports* 100: 180-188.

Smith v. Des Moines, 99 F.3d 1466 (8th Cir. 1996).

Snook, S. 1991. Low back disorders in industry. *Proceedings of the Human Factors Society 35th Annual Meeting* 35: 830-833.

Sothmann, M., Saupe, K., Jasenof, D., Blaney, J., Donahue-Fuhrman, S., and Woulfe, T. 1990. Advancing age and the cardiorespiratory stress of fire suppression: Determining a minimum standard for aerobic fitness. *Human Performance* 3: 217-236.

Sothmann, M., Gebhardt, D., Baker, T., Kastello, G., and Sheppard, V. 2004. Performance requirements of physically strenuous occupations: Validating minimum standards for muscular strength and endurance. *Ergonomics* 47: 864-875.

Stamford, B., Weltman, B., Moffatt, R.J., and Fulco, C. 1978. Status of police officers with regard to selected cardio-respiratory and body composition fitness variables. *Medicine & Science in Sports & Exercise* 10 (4): 294-297.

Strickland, M., and Petersen, S. 1999. *Analysis of predictive tests of aerobic fitness for wildland firefighters.* Paper presented at annual meeting of the Canadian Society for Exercise Physiology, Canmore, CA.

Sundet, J., Magnus, P., and Tambs, K. 1994. The heritability of maximal aerobic power: A study of Norwegian twins. *Scandinavian Journal of Medical Science in Sports* 4: 181-185.

Taylor, H.C., and Russell, J.T. 1939. The relationship of validity coefficients to the practical effectiveness of tests in selection: Discussion and tables. *Journal of Applied Psychology* 23: 565-578.

Thomas v. City of Evanston, 1983. WL 4552 (N.D. Ill. Feb. 28, 1983).

Thompson, P., Franklin, B., Balady, G., Blair, S., Corrado, D., Estes, M., Fulton, J., Gordon, N., Maron, B., Mittleman, M., Pelliccia, A., Wenger, N., Willich, S., and Costa, F. 2007. Exercise and acute cardiovascular events: Placing the risks into perspective. *Medicine and Science in Sports and Exercise,* 39:886-897.

Thompson, S., and Sharkey, B. 1966. Physiological cost and airflow resistance of respiratory protective devices. *Ergonomics* 9: 495-502.

Title I of the Civil Rights Act of 1991, Pub.L.No. 102-166, 105 Stat. 1071 (1991) (amending the CRA of 1964, 42 U.S.C.2000e et seq.).

Township of Bridgewater, N.J. v. P.B.A. Local 174, 482A.2d 183 (1984).

Uniform Guidelines on Employee Selection Procedures. 2006. 29 C.F.R. § 1607.1 et seq.

United Paramedics of L.A. v. City of L.A., 936 F2d 580 (9th Cir. 1991).

United States Fire Administration. 2002. Firefighter Fatality Retrospective Study. Washington D.C.: Federal Emergency Management Agency. www.usfa.fema.gov/dhtml/inside-usfa/fa-220.cfm/.

United States Public Health Service, Office on Smoking and Health and Office of the Surgeon General. 1985. The health consequences of smoking: Cancer and chronic lung disease in the workplace: A report of the Surgeon General. Lynn, W.R., (ed.). DHHS (PHS) 85-50207. http://profiles.nlm.nih.gov/NN/B/C/B/N.

U.S. Equal Employment Opportunity Commission. 1997. The Equal Pay Act of 1963. United States Code, vol. 29, sec. 206(d). www.eeoc.gov/policy/epa.html.

U.S. Department of Labor. 1968. *Dictionary of occupational titles.* Washington, DC: U.S. Government Printing Office.

U.S. v. City of Erie, 411 F.Supp. 2d 524 (W.D.Pa 2005).

U.S. v. City of Wichita Falls, 704 F.Supp. 709 (N.D. Texas 1988).

U.S. v. New Castle (Consent decree), (Delaware 1984).

Utica Professional Firefighters v. City of Utica, NY PERB #U-18370, 32 NYPER(LRP) P4570, 919990.

Vanderburgh, P., and Flanagan, S. 2000. The backpack run test: A model for a fair and occupationally relevant military fitness test. *Military Medicine* 165: 418-421.

Van Dorn, K. 1987. *Prediction of cycling time-trial performance from $\dot{V}O_2max$, anaerobic threshold and cycling efficiency in sitting and standing bicycle test protocols.* Unpublished master of science thesis, University of Montana.

Vogel, J.A., Wright, J.E., and Patton, J.F. 1980. *A system for establishing occupationally related gender free physical fitness standards.* U.S. Army Research Institute of Environmental Medicine. Technical Report #T 5/80, Natick, MA.

VonHeimburg, E., Rasmussen, A., and Medbo, J. 2006. Physiological responses of firefighters and performance predictors during a simulated rescue of hospital patients. *Ergonomics* 49: 111-126.

Wallace, R., Kriebel, D., Punnett, L., Wegman, D., Wenger, B., Gardner, J., and Gonzalez, R. 2005. The effects of continuous hot weather training on risk of exertional heat illness. *Medicine & Science in Sports & Exercise* 37: 84-90.

Warburton, D., Haykowsky, M., Quinney, A., Blackmore, D., Teo, K., Taylor, D., McGavock, J., and Humen, D. 2004. Blood volume expansion and cardiorespiratory function: Effects of training modality. *Medicine & Science in Sports & Exercise* 36: 991-1000.

Washburn, R., Sharkey, B., Narum, J., and Smith, M. 1982. Dryland training for cross-country skiers. *Ski Coaching* 5: 9-12.

Webster's Desk Dictionary of the English Language. 1983. New York: Portland House.

Waters, T., Putz-Anderson, V., Garg, A., and Fine, L. 1993. Revised NIOSH equation for the design and evaluation of manual lifting tasks. *Ergonomics* 36: 749-776.

Welle, S., and Thornton, C. 2002. High protein meals do not enhance myofibrillar synthesis after resistance exercise in 62- to 75-yr-old men and women. *American Journal of Physiology* 4: 677-683.

White, D. 1996. *Musculoskeletal disorders related to cigarette smoking and tobacco use.* Natick, MA: Technical report, U.S. Army Research Institute for Environmental Medicine.

White v. Village of Homewood, 628 N.W.2d 616 (Ill. App. 1 Dist. 1993).

Wilkinson, D. 2005. *The physiological loads associated with evacuation of casualties from an underground railway.* Paper presented at the annual meeting of the American College of Sports Medicine, Nashville, TN.

Williams, A., Rayson, M., and Jones, D. 2002. Resistance training and the enhancement of the gains in material-handling ability and physical fitness of British Army recruits during basic training. *Ergonomics* 45: 267-279.

Wilmore, J. 1983. *Athletic training and physical fitness.* Boston: Allyn and Bacon.

Wilmore, J., and Costill, D. 2006. *Physiology of sport and exercise.* Champaign, IL: Human Kinetics.

Wilson, J., and Raven, P. 1989. Clinical pulmonary function tests as predictors of work performance during respirator wear. *American Industrial Hygiene Association Journal* 50: 51-57.

Zamlen v. City of Cleveland, 686 F.Supp. 631 (N.D. Ohio E. Div., 1988).

Index

Note: The italicized *f* and *t* following page numbers refer to figures and tables, respectively.

A

absenteeism 51, 64, 89-90, 102, 123, 136
acclimatization, to environment 11, 141-142, 147-148
acid–base balance 117
acrolein, in smoke 156, 156*t*
actin 21, 111, 111*f*, 114*f*, 120
activity-specific training 74-75, 83
  evaluation of 87-94
  for job tasks 52, 123, 176, 183, 188
adaptations to training 119-120, 128
  General Syndrome of 182-183, 182*f*, 183*f*
adenosine triphosphate (ATP) 5, 113-116, 114*f*, 115*t*
adipose tissue. *See* body fat
adverse impact 31, 49, 54, 68, 82, 167, 169, 196
advocacy groups, for workers 25, 67, 168
aerobic capacity 14-15, 123, 169
aerobic energy 4-5, 113-116, 114*f*, 128
aerobic fitness 14, 155
  age impact on 78-79, 78*f*
  energy expenditure and 4-5
  expected rate of loss 78, 182-183, 182*f*
  factors influencing 16-20, 148, 155
  gender-based 16-17, 20, 170, 186-187, 187*f*
  in health-related fitness 104
  job-related 56, 123-126, 125*f*-126*f*
  lactate thresholds in 124-125, 125*f*
  law enforcement functions and 217-218, 218*f*
  lifting guidelines and 160-161, 164-166
  measurement of 14-15, 15*f*, 15*t*
  sustainable 125-127, 126*f*, 127*t*, 128*t*
  work implications of 7, 7*f*, 10-11, 80
aerobic fitness test
  job-related 10, 44-45, 45*f*, 51
  for military recruits 75, 186-188, 187*f*
aerobic power 15, 15*f*, 15*t*, 123
aerobic training
  benefits of 20, 78-79, 83, 118
  job-related 126-128, 127*t*, 128*t*
affirmative action/equal employment opportunity (AA/EEO) 31, 73
age/aging
  chronological *vs.* physiological 79-80
  employment opportunity issues and 24, 28, 30-31
  as fitness factor 17-18, 18*f*, 20-21
  job-related test standards and 53-54, 56, 63
Age Discrimination in Employment Act (ADEA) of 1967 28-29, 78, 168, 192
agricultural workers 4, 8, 9*t*, 24
air-purifying respirators (APRs) 153, 155, 156*f*, 157

airway anatomy, upper and lower 117, 117*f*
Alexandria Volunteer Fire Department v. Rule 170, 172
altitude, acclimatization to 72, 147-148
alveoli, physiology of 117, 117*f*
ambient temperature
  high. *See* heat disorders
  low. *See* cold conditions
  manual timber harvesting and 221*t*, 222
American College of Sports Medicine (ACSM) 60, 64, 79, 132, 147, 174
American Heart Association (AHA) 68, 134-135
American Industrial Hygiene Association (AIHA) 155
Americans with Disabilities Act (ADA) of 1990 29, 68, 168, 175, 180, 192
amino acids, in work physiology 113, 119
anaerobic energy
  aerobic energy *vs.* 4-5, 113, 114*f*, 115-116, 128
  lactate threshold and 124-125, 125*f*
  law enforcement functions and 217-218, 218*f*
1 antitrypsin, stress and 208, 209*t*
appeals 68, 81, 196-197
applicants
  medical examinations for 68-69, 70*f*, 71, 98-99
  testing of 67-76
    medical standards for 68-69, 70*f*, 71
    periodic 81
    pretest training for 73-74, 74*t*
    recruitment and 63, 72-73, 80, 174
    retesting 52-53, 74, 83
    safety considerations of 71-72
    work hardening and 74-75
  work output predictors for. *See* productivity
arbitration, on periodic testing 81-82
arm work, respirators impact on 155
athletes, fitness factors of 2, 16-17, 100
atrophy, of muscle, with age 21, 79
attitudes, employee 92
attrition 51, 90, 92, 136

B

back health
  evaluation of 83-84, 89, 99
  injury mechanisms and 38, 89, 160, 164, 182, 188
  maintenance of 104, 130
backpacks 75, 127-128, 127*t*, 160
balance, as fitness factor 51, 130
behavioral changes
  for fitness 18, 24, 39, 132, 135, 188
  in impact evaluation 88-89
benzene, in smoke 156, 156*t*

bias, in job-related tests 41, 72
bicycle test 14
biochemical evaluation 99, 103
  of firefighters 33, 42, 43*t*, 44, 47
  in job task analyses 44-45
  of workplace stress 207-209, 209*t*
biomechanical standards, for lifting 160, 166
blood glucose. *See* glucose level
blood pressure
  health screening for 89-90, 93, 103-104
  minimum qualifications for 99-100
blood tests. *See* biochemical evaluation
body composition
  fitness and 17-18, 21, 50, 105
  weight training impact on 131-132
body fat
  aerobic fitness and 17-18, 77
  as energy supply 112, 115, 115*t*, 119
body mass index (BMI) 14, 24, 50
  loss of, manual timber harvesting and 221-222, 221*t*
  in medical standards 68, 70*f*, 93, 105
body mechanics, for lifting 160, 166
body size, aerobic fitness and 18, 19*f*, 20
bona fide occupational qualification (BFOQ) 28, 53, 78
Borg scale, of exercise/exertion intensity 127-128, 128*t*
breathing zone samples, of wildland firefighters 156, 156*t*
business necessity, of tests 31, 37-39, 167, 169, 174

C

calisthenics 134
calories 119
  for weight gain/loss 131-132
  for work physiology 112-113, 142, 146
calorimetry, indirect 4-5, 7
cancers 101-104
capillaries, pulmonary, physiology of 117, 117*f*
carbohydrate intake
  environmental factors and 112, 143-144, 148, 222
  for work physiology 111-114, 114*f*
carbon dioxide, physiology of 116-117
carbon monoxide 152-153, 155-156, 156*t*
cardiac output 117-120, 118*f*
cardiorespiratory fitness. *See* aerobic fitness
cardiovascular disease
  medical examination for 100-101, 105
  physical activity risk and 38-39, 39*f*, 53, 134-135
  physically demanding jobs and 38, 77, 101-102, 152, 188
  risk factors for 89-90, 93, 118, 134-135

cardiovascular system
   minimum health for 98, 100-101
   in work physiology 117-118, 118*f*
catecholamines, stress impact on 135, 208-209
cerebrovascular disease 101
chemical/biologic/radiological/nuclear protective (CBRN) clothing 145, 152
chest X ray, in medical examination 99-100, 104
cholesterol level. *See* lipid profiles
cigarette smoking 18, 39, 89, 103, 152, 188
City of Erie, United States v. 60, 67, 76, 168, 171, 179
Civil Rights Act (CRA)
   of 1964 28, 67, 168
   of 1991 29, 173-174, 180, 192
Civil Service Commission 30, 44, 46
Civil Service Reform Act (CSRA) of 1978 29
"coasting" 183-184
cold conditions, working in 146-147, 147*f*
Combat Challenge, Firefighter 43, 145-146, 213-214
combat training, basic 8, 9*t*, 62, 123
compensation 24-25
   for fitness 84, 105, 134
   laws regarding 28, 31-32
confidentiality 61, 92
construction work 4, 7-8, 9*t*, 24
   muscular fitness for 6, 83, 164
construct validity, of job-related tests 51
content-validity
   of ability tests 186, 192-193
   of job-related tests 44-45, 49-50
contractile proteins, in muscles 21, 111, 111*f*, 113, 120, 128
coordination tests 51
core competencies, for law enforcement 176, 217-219, 218*f*
core temperature, monitoring of 144
core training 22-23, 104, 120, 130
   lifting guidelines and 164-166
coronary arteries 117-118, 118*f*
   disease of. *See* cardiovascular disease
correlation
   of job-related tests
      to job tasks 44-47, 46*t*
      to successful performance 51
   in program evaluation 91
cortisol 104, 119, 207-208, 209*t*
cost-benefit analysis
   of employee health programs 106-107, 106*f*
   of incumbent testing 82
   of job-related fitness programs 132, 135-136
   of medical clearance/examinations 100-101
   in program evaluation 92-93
court decisions
   employment regulations influencing 17, 25, 27-29, 32, 173
   on fitness requirements 17, 25, 152, 179, 181, 188
   on job-related entry/fitness standards 168-169, 171-174

on job-related tests 39-40, 45, 54-55, 60, 170-171
C-reactive protein (CRP) 89, 208, 209*t*
creatine kinase (CK), stress and 208-209, 209*t*
criminal population, analyses of 180-181
criterion-related validity
   of job-related tests 50-51
   of physical ability test 185, 185*f*, 205-206
criterion task test (CTT) 32-33, 41, 195-197
cross-validation, test reliability *vs.* 52-53
cutoff point, for hiring 169, 201-202
cutoff score, for job-related tests 55-56, 55*t*, 169
cycles
   in periodization training 131
   rest *vs.* work, for heat disorders 144

D
Davis, Paul 30, 169, 179-189, 195-197
defensive medicine 100, 106
dehydration 140, 141*f*, 142-144
demographics, as fitness factor 88, 103 23-25
Des Moines v. Civil Service Commission 170, 172
diabetes 104, 119
diet. *See* nutrition
disability
   employment opportunity issues and 28-31, 89, 98, 180, 192
   risk factors for 14, 77, 136, 152, 161
discrimination, employment 30, 32
   of firefighters 17, 195-197
   laws regarding 27-31, 167-168
diseases
   coronary. *See* cardiovascular disease
   health care costs of 90, 92, 106, 136
   health screening for 102-104
   preexisting 98-99
   pulmonary. *See* respiratory diseases
   surveillance system for 90
disparate impact. *See* adverse impact
distance 6
   in lifting equation 162, 163*f*, 164, 166
DNA, in muscle physiology 114*f*, 119
Dothard v. Rawlinson 54
Douglas bag 9-10
drug abuse 89, 104
"duck validity," of physical ability test 186, 186*f*

E
ear, nose, and throat, health of 98
Eberle v. State of Missouri Department of Corrections 172
EEOC Enforcement Guidance on Preemployment Inquiries under the ADA 172
EEOC v. Simpson Timber C. 55
80 percent rule 169-170
electrocardiogram, applications of 98-100, 134
electrolyte replacement, for heat disorders 142-143
emergency plan, for job-related tests 71, 82

emergency services. *See* public safety occupations
employee groups, average productivity of 199
employee health 97-107
   medical examinations for 99-101
   medical standards for 98
      for applicants 68-69, 70*f*, 71
      critical 99-100
      for incumbents 82
      minimum qualifications 98-99
   surveillance system for 90-92, 101-102
   task-related tests for 38-39, 39*f*
   wellness programs for 102-107
employee health programs 102-107
   costs *vs.* benefits of 96, 102, 106-107, 106*f*, 175
   evaluation of 87-94
   health education in 104
   health-related fitness in 102, 104-105
   health risk analysis in 102-104
   health screenings in 102-104
   job-related fitness in 102, 104
employee involvement
   in program evaluation 51, 90, 92-93
   in test implementation 61, 71, 80-81
employment. *See also* hiring
   Davis' 10 Axioms of 30, 179-189
employment laws 27-29, 32, 173. *See also specific law*
   court decisions and 17, 25, 39, 179, 181, 188
employment opportunity 27-34
   adverse impact and 31-32, 34, 54
   case studies of 32-33
   uniform selection guidelines for 30-32
endocrine system 99, 119
endurance 51
   musculoskeletal. *See* muscular endurance/fitness
endurance training
   adaptations in 38, 119-120
   evaluation of 87-94
   gender influence on 16-17, 21, 170
   job-related 128-130, 129*t*
   muscular factors of 21-22, 83, 119, 164
   pretest for 52, 61-62, 73, 74*t*
   tests for 22, 23*t*, 51
energy, for work physiology
   oxygen and 115-116, 116*f*, 118*f*
   pathways for use of 114-115, 115*f*, 124
   sources of 112-114, 114*f*, 115*t*
energy expenditure
   in lifting equation 161-162
   in physical work 4-5, 4*t*, 9-10
energy requirements, of work tasks 7, 8*t*
   occupational classifications for 8, 9*t*
environmental impacts 139-149
   altitude as 147-148
   cold conditions as 146-147, 147*f*
   heat disorders and 140-141
      indexes for 144
      prevention of 141-146
   heat stress as 140-141, 144*f*
   work capacity and 5, 11, 16, 123, 164
Environmental Protection Agency (EPA) 153, 157
epinephrine 119, 208-209, 209*t*

Equal Employment Opportunity
  Commission (EEOC)
    discrimination decisions 46, 168,
      196-197
    role of 28, 30, 78, 173, 196
Equal Pay Act (EPA) of 1963 28, 31-32
ergonomics 38, 44, 89, 102, 161
essential functions. *See* job task analyses
  (JTAs)
ethnicity, employment opportunity issues
  and 28, 30-31, 54, 167
evaluation report, on programs 93
evaluation working group 93
excess postexercise oxygen consumption
  (EPOC) 116, 116f
executive summary, in evaluation report
  93
exercise physiology 60, 64, 123
exertion
    heart disease risk and 134-135
    rating of perceived, for training 127-
      128, 127t, 128t
experts, subject matter 60-61, 81, 176,
  192-193
expert witnesses 168-169
expiration, measurements of 154, 154f,
  157

F
face validity, of physical ability test 186,
  186f
fast glycolytic (FG) muscle 110, 110t, 124-
  125, 125f
fast oxidative glycolytic (FOG) muscle
  110, 110t, 124-125, 125f
fast-twitch muscle fibers 21, 21f, 110-111,
  110t
fat-free mass (FFM), physical work and 6
fatigue 38
    environmental factors of 146, 148
    in foresters 207-209, 209t, 221-222,
      221t, 222f
    muscular factors of 20, 111-112, 129,
      160
fat intake, for work physiology 112-113,
  115, 128
Ferrante v. Niagara County 172
firearms, fitness for use of 176, 181, 184,
  184f
firefighters
    body size influence of 18, 19f, 20
    80 percent rule for 169-170
    fatality studies on 101-102, 136
    job-related tests for 39-40, 52, 195,
      196f, 213
      case studies of 46-47, 46t, 50, 55
      collective approach to 174-175
    job task analyses for 4-5, 8, 9t
      case study of 42-43, 43t
      criticality ratings of 219, 219f
    physical fitness policy for 80-81, 88,
      211-212
    physical fitness programs for 132,
      133f, 195-197, 196f
    respiratory protection for 5-6, 33, 50,
      145, 211
    smoke jumpers as 13, 44, 79, 91
    standard operating procedures for
      213-215

structural 4-6, 42-43
    wildland. *See* wildland firefighting
fire hose tests 46, 46t, 50
fitness ("fit")
    aerobic. *See* aerobic fitness
    decline over time 78-79, 78f, 122-123
    functional testing for 191-193, 192f
    muscular. *See* muscular fitness
    physical. *See* physical fitness *entries*
    work capacity and 5, 5f, 9-10
fitness instructors, certified 132, 135
fitness trail 132, 133f
FITSCAN Customized Individual Training
  Program 213-214
flexibility, muscular 22, 23t, 51, 130
fluid–electrolye balance, hydration and
  143-144
fluid intake
    for heat disorders 140, 141f, 142-144
    during manual timber harvesting 221-
      222, 221t
fluid loss, evaporative. *See* sweat loss
force (F)
    body size effect on 18, 19f, 20
    compressive, in lifting 160-161, 163f
    muscular fitness and 11, 73, 74t
      continuum concept of 6, 128,
        129t, 176
forced expiratory flow (FEF) 154, 154f
forced expiratory volume (FEV) 154, 154f
forced vital capacity (FVC) 154, 154f, 157
foresters
    lifting training for 165, 165f
    performance assessment of 221-222,
      222f
    stress evaluation for 207-209, 209t
formaldehyde, in smoke 156, 156t
free fatty acids (FFA), in work physiology
  115
frostbite 146-147
functional age 79-80
functional testing, for physically
  demanding jobs 191-193, 192f
Fundamental Rules of Evidence (FRE) 168

G
gastrointestinal system 99
gender
    aerobic fitness and 16-17, 20, 170,
      186-188, 187f
    air-purifying respirators and 157
    muscular fitness and 20-21
gender-based standards, for job-related
  tests 53-54, 56
gender discrimination, in employment
  28, 30-32, 167, 196-197
gender equity, in test implementation
  61-63
genetics, of fitness 16, 22, 119-120, 134
genitourinary system 99
glucose level, blood 103
    muscle function and 111-114, 114f,
      115t, 119
    stress and 207-209, 208f
glycogen metabolism
    energy production with 113-114, 114f,
      115t
    aerobic *vs.* anaerobic 115-116, 124
    fatigue and 111

"grandfathering" 63, 81
grievances, examples of 68, 81
growth hormone 119
gym-based tests, for law enforcement 185-
  186

H
hazardous material exposures 103, 144,
  152-153, 156, 156t
head and neck health, qualifications for
  98
health care costs 90, 92, 106, 136
health club contracts 132, 195
health education 104
Health Insurance Portability and
  Accountability Act (HIPPA) 61, 92
health-related fitness 83, 102, 104-105,
  175
health risk analysis (HRA) 102-104
health screenings 68-69, 102-104
health status
    of employees. *See* employee health
    job-related tests for 38-39, 39f
    physiological age and 79-80
hearing/hearing tests 98-99, 102
heart, diagram of 118, 118f
heart attacks 39, 53, 101, 118, 136, 169
    exertion and 134-135
heart disease. *See* cardiovascular disease
heart rate (HR) 117
    perceived exertion correlated to 127-
      128
    training effect on 119-120
    work capacity and 33, 162, 221
heat cramps 140
heat disorders 140-141
    prevention of 141-146
heat exhaustion 140
heat production, during work 142, 146
heat stress 140-141, 144f
    indexes for 144
heatstroke 140
heat tolerance 142, 144
height, body 14
helicopter ski guides 130
hematologic system 99, 117
hemoglobin 117, 120
heredity. *See* genetics
hierarchical model, of job tasks 185, 185f
high-efficiency particulate air filter (HEPA)
  154-155
hiring
    federal guidelines on. *See* Uniform
      Guidelines on Employee Selection
      Procedures (UGESP)
    functional testing for 191-193, 192f
    goal of 72-73
    ideal paradigm for, in law enforcement
      183, 183f
    job-related tests for
      development of 37-47
      implementation of 52-53, 59-64
      validation of 49-57
    level, to ensure job-related fitness 122-
      123, 122f
    lifting guidelines and 164-165
    physical ability testing for 44, 179-189
    work output and selectivity in 57,
      199-206

hormones
adrenocortical 104, 119, 207-209, 209*t*
overtraining and 135
in work physiology 119
Hotshots 80, 91
humidity 139-141, 141*f*, 145, 221*t*
hydration. *See* fluid intake
hypoglycemia 111-112, 207-208, 208*f*
hyponatremia, prevention of 143
hypothermia 146

I
illegal (undocumented) workers 24
immune system 38, 98, 135, 208
immunizations 102-103
impact evaluation, of programs 88-89
impairments. *See* disability
implants, medical, in employees 99
inactivity. *See* sedentary lifestyle
incentives, for health-related fitness 84,
105
incumbents
fitness training for 77, 83, 193
lifting guidelines and 164
medical examinations for 82, 98-100
testing of
adequate notice for 82
age impact on performance 78-79,
78*f*
medical standards for 82
periodic approach to 80-82
physiological age and 79-80
results and consequences of 63, 83
safety considerations of 82
inhalation, measurements of 154, 154*f*,
157
injuries
back/spine 38, 89, 159-166, 182, 188
cold conditions and 146-147
health status and 38-39, 181
lost-time 51, 64, 89, 123
overuse 135
respiratory 152-153
return-to-work testing for 64, 64*t*, 191-
193, 192*f*
in timber workers 207-209, 221
injury surveillance system 89-92
inspiration, measurements of 154, 154*f*,
157
insulin, in work physiology 119

J
job descriptions, in test development 41
job hazard analysis 71-72, 82
job-related fitness 121-136
aerobic 123-126, 125*f*-126*f*
health-related fitness *vs.* 83, 102, 104
hiring level for ensuring 122-123, 122*f*
muscular 128
job-related fitness programs
activity tasks in 52, 123, 176, 183, 188
aerobic 126-128, 127*t*, 128*t*
for applicants 73-74, 74*t*
benefits of 96, 121-122, 122*f*, 135-136
body composition and 22, 131-132
case studies of 126, 130-131, 135-136
core training 130
development of 123, 132, 134
evaluation of 87-94

facilities and site for 132, 133*f*
fire chiefs' personal narrative on 195-
197, 196*f*
guidelines for 131-132, 134
implementation of 74-75, 77, 83, 193
for incumbents 83
muscular 128-130, 129*t*
perceived exertion scale for 127-128,
128*t*
periodization of 131
for public service employees 184,
188-189
risks of 134-135
job-related tests 37-47
administration of 59-64. *See also* test
implementation
development of 40-47. *See also* test
development
for employee health 38-39, 39*f*
evaluation of 87-94
for firefighters 39-40, 52, 175, 195,
196*f*, 213-215
case studies of 46-47, 46*t*, 50, 55
for incumbent employees 77-85
job task analyses for 40-44
for law enforcement 39-40, 51, 175,
179-189
legal issues of 39-40, 167-171, 177
necessity of 31, 37-39, 167
for new employees 67-76
standards for 53-57, 63, 175
training and practicing impact on 52-
53, 83, 176
types of 44-45
validation of 49-57. *See also* test
validity/validation
job satisfaction 92
job task analyses (JTAs)
comprehensive taxonomy of 185, 185*f*
for firefighters 4-5, 8, 9*t*
case studies of 42-43, 43*t*, 212
criticality ratings of 219, 219*f*
for law enforcement 176, 179-180
criticality ratings of 217-219, 218*f*
for physically demanding work 44,
192-193
for screening procedures 205-206
for test development 40-44, 49-50,
52, 174

K
kilocalories. *See* calories
Klamath Falls Fire District v. KF Firefighters
170, 172
knowledge, skills, and abilities (KSAs),
job-related
adaptation of 182-183, 183*f*
decline of 78-79, 78*f*, 122-123, 122*f*
enhancement of 121, 123
essential. *See* job task analyses (JTAs)

L
laboratory tests. *See* biochemical
evaluation
labor-management agreements 61, 78, 81,
83, 105, 177
lactate/lactic acid, blood
as effort indicator 33, 51, 124
physiology of 4-5, 111, 116

response in firefighters 33, 124-125
sustainable fitness and 125-127, 126*f*,
127*t*
thresholds of 124-125, 125*f*
ladder tests 18, 19*f*, 20, 50
Lanning v. SEPTA 55, 84, 90, 168-169,
171-173, 179, 181, 188
law(s)
employment. *See* employment laws
types of 173
law enforcement
functional testing for 191-193, 192*f*
job-related tests for 39-40, 51, 174-
175, 179
gender equity and 62-63
job task analyses for 176, 179-180,
217-219, 218*f*
physical ability testing for 18, 76,
179-189
work requirements of 4-6, 8, 9*t*
lean body weight (LBW)
gender differences in 17, 21
weight training and 131-132
work capacity and 6-7, 11, 91
legal issues 167-177. *See also* litigation
80 percent rule 169-170
consensus on 176-177
of employment opportunity 27-29,
32, 168
experts for 168-169, 176
of job-related tests 39-40, 45, 54-55,
170-171
implementation considerations
60-61, 82
physical fitness as 17, 25, 152, 179,
181, 188
of preemployment testing 167-170
Legault v. aRusso 172
lifestyle decisions. *See* behavioral changes
lifting capacity
gender-based statistics on 186-187,
187*f*
repetitive 161
lifting guidelines 160-161
NIOSH equation for 161-162, 163*f*,
164
training programs and 7, 119, 164-166
worker selection and 164-166
lifting index (LI) 164
linear compensatory model, of applicant
selection 204-205
line dig test 46, 46*t*
lipid profiles, monitoring of 89-91, 93,
99, 103
litigation
alternatives to 174-176
costs of 172-174
decisions resulting from. *See* court
decisions
load(s)
aerobic capacity related to 123-124
compressive, lifting and 160-161, 163*f*
external, body size effect on 18, 19*f*,
20
in lifting equation 162, 163*f*, 164, 166
in muscular fitness training 7, 73, 74*t*,
131
in physical work requirements 6, 6*t*,
119, 164, 207

standards for. *See* lifting guidelines
in sustainable training 127-128, 127*t*
lotteries, for applicant selection 72-73
lung function. *See* respiratory *entries*

M
management/managers, involvement of 61, 63, 93
material handling 24, 73, 83
maximal heart rate (HRmax), for training 120
maximal oxygen intake (VO2 max)
    age impact on 78-79, 78*f*
    cardiac output relationship to 117, 118*f*, 120
    factors influencing 16-20, 170
    lactate threshold and 124-125, 125*f*
    physiology of 14-15, 15*f*, 15*t*
    respirators effect on 155, 156*f*
    response in firefighters 33, 42, 43*t*, 44, 47, 80, 88
    sustainable fitness and 56, 126-127, 127*t*
maximal voluntary ventilation (MVV) 116, 154-155, 157
maximum acceptable weight (MAW), for work tasks 160
mechanization, impact on muscular work 1-2
medical clearance/examinations
    cost *vs.* benefit of 100-101
    for firefighters 38-39, 53, 212, 214-215
    objective measures of 99-100, 131
        for incumbents 82
        for new employees 68-69, 70*f*, 71, 98-99
    practical applications of 38, 39*f*, 53, 212, 214-215
    for respirator use 154-155, 154*f*
medical conditions
    health screening for 102-104
    incumbent testing and 83-84
    physical fitness impact on 135-136
    preexisting 98-99
medical emergencies
    heart attacks as 39, 53, 101, 118, 136, 169
    heat disorders as 140, 145-146
    during job-related tests 71, 145-146
medical review officer 98
medical standards
    for employee health 98
        critical 99-100
        minimum qualifications 98-99, 102
    for incumbents 82
    for new employees 68-69, 70*f*, 71
medical surveillance system 90, 101-102
medications, employee, evaluation of 98, 101
mental health, qualifications for 83, 99
mentoring programs 72, 82
Meridian v. Firefighters Association of Michigan 172
metabolic system 99
    energy production by, oxidative *vs.* nonoxidative 4-5, 115-116
    muscular fatigue and 111-112

military
    basic combat training for 8, 9*t*, 62, 123
    body weight/mass of of 7, 50
    elite/special forces of 13, 80, 181
    fitness training units for 126
    lifting standards for 73, 160
    physical ability/fitness tests for 38, 52, 62, 75, 169, 186-188, 187*f*
    strength training for 50-51, 135
    work requirements of 4-5, 8, 9*t*, 28, 73, 124
military occupational specialty (MOS) 186-187
minimum qualifications/standards
    for employee health 98-99
    for job-related fitness 121-122, 122*f*, 132, 134, 136
    for job-related tests 54-55, 170-171, 182-183, 183*f*
mining, work requirements of 4, 8, 9*t*
Minnesota Department of Human Relations v. City of St. Paul 174
mitochondria, in work physiology 114*f*, 128
mixed continuous scale strategy, for applicant selection 203-204, 204*t*
modeling, of job tasks 185-168
morale, employee 90, 92-93
mortality, trends of 101-102, 106, 136
motivation
    for fitness 84, 105, 134
    work capacity and 11, 91
motor cortex, muscle contraction role of 110-111, 110*f*
motor unit 111
movement quality, in job-related tests 22, 23*t*, 51
muscle contractions, physiology of 110-111, 110*f*-111*f*, 113
muscle fibers/tissue
    characteristics and structure of 110, 110*t*, 114, 114*f*
    influences on 17, 20-21, 79
    lactate threshold and 124-125, 125*f*
    types of 21, 21*f*, 110
muscles
    cramps in, heat-related 140
    fatigue related to 20, 111-112, 160
    wasting of, cortisol effect on 208
muscular endurance/fitness 5, 22
    components of 5-6, 104
    decline of with age 79, 123
    factors influencing 20-23, 79, 164, 169
    in health-related fitness 104
    job-related 56, 128
    law enforcement functions and 176, 218, 218*f*
    lifting and 160-161, 164-166
    training for. *See* endurance training
    work capacity and 6-7, 7*f*, 11, 20
muscular fitness tests 22, 23*t*, 31, 44, 51, 176
muscular requirements, of physical work 7, 7*f*, 20
    occupational classifications for 8, 9*t*
musculoskeletal system 83, 89, 99
myocardial infarction. *See* heart attacks
myosin 21-22, 111, 111*f*, 114*f*, 120

N
National Collegiate Athletic Association (NCAA) 91-92
National Fire Protection Association (NFPA)
    essential job functions per 52, 153, 174
    fitness recommendations of 38, 106, 195, 212
    test development role of 174-176
National Institute for Occupational Safety and Health (NIOSH) 153, 161-162, 163*f*, 164
National Wildfire Coordinating Group (NWCG) 10, 52
nervous system 99
    muscle function and 21, 110-111, 110*f*
    sympathetic, stress and 208-209, 209*t*
new-hires. *See* applicants
nonoxidative metabolism. *See* lactate/ lactic acid
norepinephrine 119, 208, 209*t*
nutrition, for work physiology 112-115, 114*f*, 115*t*
nutrition programs 82-83, 104

O
obesity
    body mass index and 68, 70*f*
    fitness and 14, 18, 24, 161
    in law enforcement officers 188-189
    weight control programs for 82-83, 104, 132
objectivity
    of physical ability test 184, 184*f*
    of program evaluation 92-93
observation, for test development 41
occupational medicine 60, 64
Occupational Safety and Health Administration (OSHA) 37, 53, 99, 136, 173
    respiratory protection program of 151, 153-154, 156, 158
occupations
    physically demanding. *See* physical work
    work classifications of 8, 9*t*
    work task energy requirements for 7, 8*t*
Office of Personnel Management 30, 44, 46, 212
oral health, qualifications for 98
organic vapor/acid gas absorbent (OV/AG) 154-155
organ transplants, in employees 99
orientation sessions, for applicants 72
orthopedic conditions, evaluation of 83-84, 99
outcome evaluation, of programs 90
overload, in training effect 119, 128-129
overtraining 135
oxidative metabolism 16, 119, 148
    energy production with 4-5, 115-116
    for muscle contractions 113-116, 114*f*-115*f*, 115*t*, 128
oxygen debt/deficit 116, 116*f*
oxygen intake
    maximal. *See* maximal oxygen intake (VO2 max)
    in work physiology 115-116, 128

oxygen supply 115
  altitude impact on 147-148
  energy production and 115-118, 116*f*,
    118*f*
oxygen transport
  hazards to 18, 39, 152, 155, 156*f*, 188
  physiology of 115, 117-118, 117*f*-118*f*

P
pack test 46-47, 46*t*, 52, 64, 72, 80
  respirators effect on 155, 157
Parker v. Washington, D.C. 39-40
particulate matter (PM), in smoke 153,
    156-158, 156*t*
pass/fail strategy
  for applicant selection 50, 203-204,
    204*t*
  for job-related tests 55-56, 55*t*
passing score, minimal, for job-related
    tests 54-55
*Patrolmen's Benevolent Association v.
    Township of East Brunswick* 172
peak expiratory flow rate (PEFR) 154,
    154*f*
peak inspiratory flow rate (PIFR) 154,
    154*f*, 157
*Pentagon Force Protection Agency v. Fraternal
    Order of Police DPS Labor committee*
    172
perceived exertion, rating of, for training
    127-128, 127*t*, 128*t*
performance assessment
  during manual timber harvesting 221-
    222, 222*f*
  physiological age and 79-80
  in program evaluation 90-92
  respirators effect on 155, 156*f*, 157
  work output as. *See* productivity
performance increment scale, expected
    203-204, 204*t*
performance standards
  job-related test validity and 50-51, 167
  for physical fitness
    job-related 44-45, 121-122, 122*f*,
      132, 134, 136
    minimum acceptable 54-55, 121,
      182-183, 182*f*-183*f*
    unintended consequences of
      minimum 170-171
    value of, hiring selectivity and 193,
      200, 200*t*
periodic testing 80-83
periodization, of job-related training 131
personal protective equipment (PPE) 6,
    33, 50, 144-145, 152, 211
phosphocreatine (PCr), in muscle
    physiology 113, 114*f*, 115-116, 115*t*
physical ability testing (PAT)
  80 percent rule for 169-170
  for firefighters 50, 195, 196*f*, 214-215
  for law enforcement 30, 76, 175, 179-
    189
physical activity
  benefits *vs.* risks of 106, 106*f*, 135-136
  heart disease risk and 38-39, 39*f*, 53,
    134
  leisure-time 89, 104-105
physical characteristics, of workers 14

physical examination. *See* medical
    clearance/examinations
physical fitness
  General Adaptation Syndrome of 182-
    183, 182*f*-183*f*
  policy on 211-212
physical fitness programs. *See* job-related
    fitness programs
physical fitness tests 44-45
  aerobic. *See* aerobic fitness
  evaluation of 87-94
  for firefighters 42-43, 43*t*, 50
  for incumbent employees 77-85
  for military 38, 62, 75, 169
  for new employees 67-76
  strength. *See* muscular fitness
physical work
  capacity for 9, 11. *See also* work
    capacity
  case study of 9-10
  components of hard 4-8
  energy expenditure classifications for
    4-5, 4*t*
  energy requirements of 7, 8*t*
  fitness and 5-7, 5*f*, 10
  functional testing for 191-193, 192*f*
  lean body weight and 6-7
  mechanization impact on 1-2
  muscular requirements of 7, 7*f*
  muscular strength classifications for
    6, 6*t*
  occupational classifications of 2-4, 8,
    9*t*, 11
  physical ability testing for 179-189
physicians, employee health role of 98
physiological capacity 79-80
physiological response
  in firefighters 33, 42, 43*t*, 44, 47, 80,
    88
  to overtraining 135
physiological standards, for lifting 160-
    191
physiological strain index (PSI) 144-145
physiology of work 103-120
  energy for
    oxygen and 115-116, 116*f*, 118*f*
    pathways for use of 114-115, 115*f*
    sources of 112-114, 114*f*, 115*t*
  in firefighters 33, 42, 43*t*, 44, 47
  job-related tests and 44-45, 56, 167,
    171, 176
  muscle function and 110-112, 110*f*-
    111*f*, 110*t*, 114*f*
  stress and 207-209, 209*t*, 221
  supply and support systems for 116-
    120, 117*f*-118*f*
Pierce v. Franklin Electric Co. 172
pollution, as occupational hazard 103,
    152
potassium replacement, for heat disorders
    142-143
power, fitness aspects of 6, 22, 130, 176
PowerPoint presentation, of evaluation
    report 93
practice opportunities, for job-related tests
    52, 61-63, 82-83
preemployment testing. *See* job-related
    tests

pregnancy, health status during 99, 104
prehire selection procedure 200-201
pretest training, for job-related tests 52,
    61-63, 73-74, 82
preventive health programs, primary *vs.*
    secondary 102-104, 123
process evaluation, of programs 88
productivity
  average, of employee groups 199
  expected increase in, for selected
    applicants 200, 200*t*
  hiring selectivity and 199-206
  job-related tests for 39, 51
  manual timber harvesting and 222,
    222*f*
  predicted variation of, in unselected
    applicants 199-200
  in program evaluation 90, 92
  relative, estimating 199
  screening for. *See* screening procedures
  sustainable fitness and 9, 125-128,
    126*f*, 127*t*, 128*t*
program evaluation 87-94
  analysis of
    cost-benefit 92-93
    quantitative *vs.* qualitative 91-92
  case studies of 88-91
  impact focus of 88-89
  outcome focus of 90
  planning and design for 88
  process focus of 88
  reporting results of 93
  surveillance system for 90-92
prosthetics, in employees 99
protective clothing. *See* personal protective
    equipment (PPE)
protein intake, for work physiology 113,
    119
psychological conditions 83, 99
psychophysical standards, for lifting 160
psychosocial variables, of fitness 89, 135
public safety occupations
  fitness requirements for 71, 80, 83,
    105, 138
  functional testing for 191-193, 192*f*
  injuries of 181-182, 188
  job task analyses for 180
  physical ability testing for 18, 179-189
  standard operating procedures for
    213-215
pulmonary function tests 101-102, 116,
    154-155, 154*f*, 157
pulmonary system. *See* respiratory system

Q
qualitative evaluation, of programs 91-92
quantitative evaluation, of programs 91-
    92
questionnaires
  for health screening 68-69
  for program evaluation 88, 92-93

R
race, employment opportunity issues and
    28, 30-31, 54, 167
rating of perceived exertion (RPE), for
    training 127-128, 127*t*, 128*t*
recreational activities 89, 104-105

recruitment, efficient advertisements for 72-73
recruits. *See* applicants; military
red blood cells 16, 117
regulations, employment. *See* employment laws
rehabilitation
    for firefighters 64, 64t, 215
    for minimum job performance 123, 183-184, 183f
Rehabilitation Act of 1973 29
rehydration, monitoring of 143-144
reliability
    cross-validation and 52-53
    of physical ability test 184, 184f
religion, employment opportunity issues and 28, 30-31
remedial training, for job tasks 183, 188
repetitions, in muscular fitness training 73-74, 74t
    job-related 128-129, 129t
rescue services personnel. *See* public safety occupations
resistance training. *See* strength training
respirators
    air-purifying 153, 155, 156f, 157
    medical evaluation for 154-155, 154f
    OSHA standards for 153-154
    selection of 145, 152-153
    self-contained. *See* self-contained breathing apparatuses (SCBA)
respiratory diseases 101-102, 152
respiratory hazards 18, 39, 103, 152, 156t, 188
respiratory protection
    cold air and 146-147
    equipment for 152-153, 157
    indications for 145, 152
    OSHA program for 151, 153-154
    work performance and 155, 156f, 157
respiratory rate 116
respiratory system 98
    function testing for 101-102, 154-155, 154f, 157
    in work physiology 116-117, 117f
retesting 52-53, 74, 83
return-to-work, testing for 64, 191-193, 192f
rewards, for health-related fitness 84, 105
rhabdomyolysis, exertional 141
risk-benefit analysis, of job-related fitness programs 134-136
RNA, in muscle physiology 114f, 119

S
safety, job-related tests and 37-38, 71-72, 82
salt replacement, for heat disorders 142-143
sarcopenia 79
scores, for job-related tests 54-57, 55t, 169
screening procedures, for work output 199-206
    concluding observations on 205-206
    expected performance increment scale 203-204, 204t
    information loss and 202-203, 203t
    limitations of 202

lower-bound utility with predictor zero weighting 204-205
    mixed continuous scale strategy 203-204, 204t
    multiple predictors in 202
    pass/fail strategy 203-204, 204t
    prehire use of 200-201
    suboptimal 57, 205-206
    top-down hiring *vs.* cutoff point 201-202
    utility of 201-202
    validity of 201-203, 203t, 205
sedentary lifestyle 68-69, 77-78, 116, 118
    health risks of 24, 39, 134-135
self-contained breathing apparatuses (SCBA)
    impact on work capacity 5-6, 33, 50, 211
    indications for 145, 153
SEPTA
    Lanning v. 55, 84, 90, 168-169, 171-173, 179, 181, 188
    performance assessment at 90, 92
short stature, aerobic fitness and 18, 19f, 20
shovel test 46, 46t
simulation, of job tasks 185-168
skin, health of 98
slow oxidative (SO) muscle 110, 110t, 124-125, 125f
slow-twitch muscle fibers 21, 21f, 110-111, 110t
Smith v. Des Moines 172
smoke exposures
    cigarettes as 18, 39, 89, 103, 152, 188
    occupational 152-153, 156, 156t
smoke jumpers 13, 44, 79, 91
smoking cessation programs 89, 104
social engineering 171
Society for Industrial Organizational Psychology 57, 64
sodium intake, for heat disorders 142-143
sodium level, plasma, hydration effect on 143-144
software, for program evaluation 91
special weapons and tactics (SWAT) team 181
specificity, in training 126, 128
speed, as power factor 6
spinal cord, muscle contraction role of 110-111, 110f
spine
    compression, with lifting 160-161, 163f
    injuries of 38, 160, 164, 182, 188
sport drinks, for heat disorders 143-144
spreadsheets, for program evaluation 91-92
stability, core, factors influencing 22, 130
stadiometry, precision 161
stair climb test 14, 50
stamina tests 51
standard operating procedures, for firefighters 213-215
statistical analysis, in program evaluation 91-93
step test, five-minute 80
strength, muscular 5, 20

factors influencing 20-23, 79, 123
    gender-based statistics on 169, 186-187, 187f
    law enforcement functions and 176, 218, 218f
    lifting and 160-161
    tests for 22, 23t, 31
    training for 7, 38, 52, 61-62, 83
    work classifications for 6, 6t
    work rate and 7, 7f
strength-endurance continuum 128, 129t
strength training
    benefits of 21-22, 79, 83, 120
    evaluation of 87-94
    job-related goals for 128-130, 129t
    lifting guidelines and 164-166
    for military 50-52
    pretest for 52, 61-62, 73-74, 74t
stress, workplace 104, 119, 136
    biochemical evaluation of 207-209, 209t
stress management 104
stress test. *See* treadmill test
stroke volume (SV) 117, 119-120
subjectivity, of program evaluation 92-93
subject matter experts (SMEs) 60-61, 81, 176, 192-193
substance abuse 89, 104
sudden death 39, 53, 100, 102, 136
    exertion and 134-135
surveillance system, for employee health 90-92, 101-102
survey instruments, for test development 41, 43
sustainable fitness 9, 125-128, 126f, 127t-128t
sweat loss
    heat disorders and 140-142, 144
    during manual timber harvesting 221, 221t

T
talk scale, of perceived exertion 128t
target practice, as test 184, 184f
tasks. *See* work tasks
taxonomy, comprehensive, of job tasks 185, 185f
technology, impact on muscular work 1-2
teen employment 24
temperature
    air. *See* ambient temperature
    core, monitoring of 144
test development 40-47
    collective approach to 174-175
    consensus approach to 176-177
    criteria in 45
    independent experts for 81, 176
    predictive validity factors of 44-45
    strategic tools for 40-44
test implementation 59-64
    for incumbent employees 77-85
    for new employees 67-76
    periodic, options for 80-83
    personnel issues of 61-63, 71
    professional assistance for 60
    strategies for 63-64
testosterone, overtraining and 135
test–retest consistency 52-53

tests
administration/implementation of
52-53, 59-64
development of 40-47
evaluation of 87-94
functional 191-193, 192f
job-related. See job-related tests
physical ability. See physical ability
testing (PAT)
physical fitness. See physical fitness
tests
standards for 53-57, 63, 169-171,
182-183
validation of 49-57
test standards
absolute vs. relative 53-54, 63
cutoff score and 55-56, 55t, 169
minimum job-related 54-55, 170-171,
182-183, 183f
ranking of 56
work setting scores and 56-57
test validity/validation 49-57
collective approach to 174-175
construct 51
content 49-50, 186, 192-193
criterion-related 50-51, 185, 185f,
205-206
"duck" 186, 186f
of functional tests 192-193
of job-related tests 44-45
components of 49-53
independent experts for 176
standards for 28, 53-57, 63
options for 51-52, 57
of physical ability tests 184, 184f,
186-188
reliability vs. 52-53
suboptimal selection procedures
despite 57, 205-206
think scale, of perceived exertion 128t
Thomas v. City of Evanston 170, 172
throw test 46, 46t
tidal volume (TV) 116
timber harvesting 221-222, 222f
time/distance factor 6
in sustainable training 126-127, 127t,
128t
time-share, for job-related training 134
top-down approach, to hiring 201-202
Township of Bridgewater, NJ. v. P.B.A. Local
174 172
training effect 119
training facilities/site, for job-related
programs 132, 133f
training intensity 127-128, 127t, 128t
training programs
aerobic 20, 78-79, 83
core 22-23, 104, 164-166
endurance. See endurance training
evaluation of 87-94
for job-related fitness 121-136
for job-related tests 52, 61-63, 73-74,
82-83, 90, 183
muscular. See strength training
treadmill test
for aerobic fitness 14, 50, 53, 157
exertional risks with 134-135

lactate threshold and 124-125, 125f
in medical examination 100-101, 104,
155
tree planting 165, 165f, 207-209, 209t
triglycerides, in muscle physiology 115
turnover, employee 51, 90, 92

U
Uniform Guidelines on Employee
Selection Procedures (UGESP)
adverse impact and 31-32, 34
discrimination and 30-31, 167
purpose of 28, 30, 44, 173, 186, 192
test validation approaches per 49,
55, 57
unions, role of 61, 78, 81, 83, 105, 177
United Paramedics of LA v. City of Los
Angeles 172
United States v. City of Erie 60, 67, 76, 168,
171, 179
United States v. City of Wichita Falls 172
United States v. New Castle 172
urine output
heat disorders and 140, 142-143
during manual timber harvesting 221
U.S. Congress, authority of 173
U.S. Department of Defense 8, 9t, 62, 123
U.S. Department of Energy 81
U.S. Department of Justice 68, 168, 173
U.S. Department of Labor 173
Utica Professional Firefighters v. City of Utica
172

V
validity/validation
of screening procedures 201-203, 203t,
205
of tests 49-57. See also test validity/
validation
variance, explained, of job tasks 185, 185f
ventilation. See pulmonary function tests
vision/vision test, qualifications for 98,
100

W
wages. See compensation
walking, as exercise 68-69
warm up, importance of 134
water loss, evaporative
heat disorders and 140-144
during manual timber harvesting 221,
221t
weapons, fitness for use of 176, 181, 184,
184f
weight, body 14, 24
excess. See obesity
fitness and 18, 19f, 131-132
health screening for 89, 91, 103
lean. See lean body weight (LBW)
weight loss
monitoring, for dehydration 144
programs for 82-83, 104, 132
weight standards
for backpacks 75, 160
limits, in lifting equation 161-162,
164, 166
weight training. See strength training
wellness programs. See employee health
programs

wet bulb globe temperature (WBGT) 144
white blood cells (WBC), stress and 207-
208
White v. Village of Homewood 172
wildland firefighting
breathing zone samples of 156, 156t
fitness standards for 17, 52, 71, 81, 91
fitness training facilities for 132, 133f
job task analysis for 42-43, 46, 46t
medical standards for 7, 79, 98
work capacity for 9-10, 64, 64t, 91,
176
work requirements of 4, 8, 9t
wind chill 146, 147f
work capacity 9
altitude impact on 147-148
factors of 5, 5f, 10-11, 91
maintenance of 167, 170, 175. See also
job-related fitness programs
testing for 72-73
consequences for incumbents 83-
84
evaluation of 87-94
in wildland firefighters 9-10, 64,
64t
workers 13-25
aerobic fitness factors in 14-20
aerobic training for
benefits of 20, 78-79, 83, 118
job-related 126-128, 127t, 128t
core training for 22-23, 104, 120, 130
demographic availability of 23-25
muscular fitness in 16-17, 20-23, 23t,
51, 73, 170
physical characteristics of 14
strength training for 21-22, 52, 61-62
job-related 128-130, 129t
workers' compensation costs 51, 90, 102,
130
work hardening 74-75
for job tasks 52, 176, 183, 188
work output. See productivity
work performance. See performance
assessment
work physiology. See physiology of work
workplace setting 137-138
environmental impact issues of 139-
149
legal issues related to 138, 152, 167-
177
lifting guidelines based on 159-166
performance scores based on 56-57
respiratory protection and 151-158
workplace stress 104, 119, 136
biochemical evaluation of 207-209,
209t
work rate, physical factors of 7, 7f
work tasks
analysis of. See job task analyses (JTAs)
criterion test of 32-33, 41, 195-197
energy requirements for 7, 8t
in lifting equation 162, 163f, 164
physical. See physical work
World Health Organization 103

Z
Zamlen v. City of Cleveland 172

About the Authors

Brian J. Sharkey, PhD, is a physiologist in the Technology and Development Center at the United States Department of Agriculture (USDA) Forest Service in Missoula, Montana, where he researches fitness, health, and work capacity. Previously, Sharkey served as director of the University of Montana's Human Performance Laboratory and remains associated with the university and lab as professor emeritus.

Courtesy of Brian Sharkey

Paul O. Davis, III, PhD, is the president of the First Responder Institute in Burtonsville, Maryland, where he has conducted job and medical standards development for hundreds of public safety and military organizations. He is a former firefighter/paramedic and as a member of the Fire Board of Montgomery County responsible for the development of definitive medical care outside of the hospital.

As a leading fitness researcher, educator, and author, Sharkey has more than 40 years of experience in both exercise and work physiology, including research with wildland firefighters. For contributions to the health, safety, and performance of firefighters, Sharkey received the USDA's Superior Service Award in 1977 and its Distinguished Service Award in 1993.

Sharkey is a past president of the American College of Sports Medicine and served on the NCAA committee on competitive safeguards and medical aspects of sports, where he chaired the Sports Science and Safety subcommittee, which uses research and data on injury to improve the safety of intercollegiate athletes. He also coordinated the United States Ski Team Nordic Sports Medicine Council.

In his leisure time, Sharkey enjoys cross-country and alpine skiing, road and mountain biking, running, hiking, and canoeing. He lives near his grandchildren in Missoula, Montana.

As an expert witness, Davis has made more than 60 appearances in federal and state court and was recruited by the FBI to participate in legal defense of physical standards. He was also selected by the United States Marine Corps to validate the physical fitness test (PFT) and to conduct certification of the physical training unit staff at the FBI Academy at Quantico, Virginia. Most recently he was engaged by the Department of Homeland Defense to develop hiring and retention standards for the reorganized Immigration and Customs Enforcement (ICE-D). He is the creator of several TV sports productions including the Firefighter Combat Challenge providing color commentary on ESPN, A&E, and the Versus network.

Davis is a fellow of the American College of Sports Medicine. He received his PhD in exercise science in 1976 from the University of Maryland.